HOW SIMPLE THINGS ARE MADE

HOW SIMPLE THINGS ARE MADE

by Danelle McCafferty

Illustrations by
Michael Goodman

Photographs by
Sam and Stephen Antupit

SUBSISTENCE PRESS

A PLUME BOOK
NEW AMERICAN LIBRARY

TIMES MIRROR
NEW YORK, LONDON, AND SCARBOROUGH, ONTARIO

Copyright © 1977 by Subsistence Press, Inc.
All rights reserved.

Library of Congress Catalog Number: 76-56048

ISBN: 0-452-25144-3

Design by Antupit and Others
Published by arrangement with Subsistence Press, Inc.

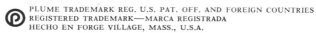 PLUME TRADEMARK REG. U.S. PAT. OFF. AND FOREIGN COUNTRIES
REGISTERED TRADEMARK—MARCA REGISTRADA
HECHO EN FORGE VILLAGE, MASS., U.S.A.

SIGNET, SIGNET CLASSICS, MENTOR, PLUME and MERIDIAN BOOKS
are published *in the United States* by
The New American Library, Inc.,
1301 Avenue of the Americas, New York, New York 10019,
in Canada by The New American Library of Canada Limited,
81 Mack Avenue, Scarborough, 704, Ontario,
in the United Kingdom by The New England Library Limited,
Barnard's Inn, Holborn, London, E. C. 1, England.

First Plume Printing, March, 1977

1 2 3 4 5 6 7 8 9

PRINTED IN THE UNITED STATES OF AMERICA

To my children,

Lisa and Amy Ucko

I wish to thank the following for their help in supplying information and materials:

American Iron and Steel Institute; American Paper Institute; Berol Corporation; B. F. Goodrich Engineered Systems Company; Binney & Smith, Inc., maker of CRAYOLA® crayons; Brick Institute of America; Buglecraft, Inc.; Chattanooga Glass; *Cocoa*, courtesy of Merrill, Lynch, Pierce, Fenner and Smith; Colonial Mirror & Glass Corp.; Columbia Records; *Correctional Services News*; Diamond International Corporation, Diamond Match Division; L. A. Dreyfus Company; Frito-Lay, Inc.; The Goodyear Tire & Rubber Company; Hershey Foods Corporation; International Paper Company; Kellogg Company; International Association of Ice Cream Manufacturers; Johnson & Johnson; Keds Sneakers by Uniroyal, Inc.; Levi Strauss & Company; Maple City Rubber Company; Marble King, Inc.; National Association of Chewing Gum Manufacturers; New York State Department of Correctional Services; Oneida Ltd., Silversmiths; Oscar Mayer & Company; Official Baseball Rules, The Sporting News; Bill Pestone; PPG Industries; The Proctor & Gamble Company; Recording Industry Association of America, Inc.; Leonard Schlosser; Sewing Notions Division, Scovill Manufacturing Company; The Society of the Plastics Industry, Inc.; Tek Hughes Division, International Playtex, Inc.; Union Camp Company; Union Carbide Corp., Films Packaging Division; U.S. Steel; U.S. Treasury Department; Wilson Sporting Goods Company; Wallace Pencil Company; Josiah Wedgewood & Sons, Ltd.; Wrangler® Jeans and Sportswear, manufactured by Blue Bell, Inc.

And special thanks to Robert Abel, Sam Antupit, Richard Baron, Valerie Dow, Harold Messing, Randall Swatek, and John Thornton, without whom this book could not have been made.

CONTENTS

How are simple things made? Well, not so simply, as it turns out.

Simple things—meaning familiar materials and objects that we see and use frequently—seem to fall into two categories. First, there are objects such as soap or bricks which for centuries man made simply, by hand, but which today are manufactured in great quantity with sophisticated machinery. Second, there are products of the modern age which have never been made simply such as records, toothbrushes, or tires, but which seem simple because they are so common.

In fact, everything in this book is made by machine. Some things, like marbles or chewing gum, are produced from start to finish without ever being handled. Others, like blue jeans, are put together by many people, assembly-line fashion.

The purpose of this book, then, is to explain simply and clearly, step-by-step, how some so-called simple things are made.

In our research a number of interesting facts turned up. We found unusual jobs such as the whistle tester who listens for defective whistles, unusual machines such as the picker-sheller, little known history—Indians in Brazil, for instance, were making their own sneakers out of rubber latex 300 years ago, and some unusual uses for common objects, such as using marbles to roll caskets into mausoleums and mix paint in aerosol cans.

We found ourselves getting carried away a bit unraveling baseballs, pulling bristles out of toothbrushes, shredding adhesive bandages and sampling different ice creams to test our information.

As the book progressed, it became evident that once some basic processes are known, it's not hard to understand how products are made from there—how glass becomes a bottle, how steel becomes a safety pin, how rubber becomes a balloon, how plastic becomes a straw. And perhaps more important, once it's known how some simple things are made, it's not hard to figure out how other things are made as well.

WOOD

MATCHES

The American Indians used a simple drill to start their fires. It had two pieces, a round pointed stick and a flat piece of wood with a groove in it. They put dry leaves and bits of wood and twigs around the groove and then placed the tip of the stick in the groove. They twirled the stick between the palms of their hands. The friction of the stick rubbing in the hole on the board made heat, and after a while the leaves and twigs began to burn.

Flint was used by the pioneers to make fire. Flint is a type of quartz, a hard rock, which causes sparks when struck against steel.

In 1669 an alchemist named Henry Brandt accidentally discovered the chemical element phosphorus while he was trying to find a way to make gold. Since phosphorus explodes into flames very easily it became popular for making fire.

The first match was made in 1680. The chemical sulfur was put on the tips of small pieces of wood. The splinters were pulled through a folded, rough piece of paper coated with the chemical phosphorus. Friction caused the splinter to light. But these matches cost a lot of money (an ounce of phosphorus cost the equivalent of $250 today), and they were very

8.

dangerous, so people stopped using them and simply continued making fires by striking flint against steel.

It took about 150 more years to develop the kinds of matches we now use. Today we have the phosphorus-tipped wood match which can be lit by striking it against anything rough. Then there is the safety wood match which will only light when struck against the box; that's because some of the chemicals needed to make it ignite are on the package while the rest are in the match. They are sometimes called kitchen safety matches. And, of course, there are paper book matches, which are sold and given away with advertisements right on the covers.

They are all made in much the same way. The wood or paper for the splints, or body of the match, must be cut to proper size, dipped in a number of mixtures to make the heads, and dried, and the finished matches must be put in boxes or books.

The wood match is made, from beginning to end, in one machine. The machine is 60 feet long and as high as a two-story house. Blocks of pine are fed into the head of the machine. The machine cuts the blocks into match-sized strips and then clamps the bottoms of them to metal plates which are on a moving conveyor above a number of vats. The splints begin a "roller coaster" ride during which they are dipped into five different "baths" containing chemicals or wax.

First the splints are dipped in a solution of ammonium phosphate, which prevents "afterglow"—that is, it keeps the wood from smoking once the flame has gone out. Because of this, a thrown-away match is less likely to start a fire.

The splints ride to the next vat, where their tips are dipped into paraffin wax. This will help them burn more quickly.

Then the tips are dipped into chemicals and other materials which give them their heads. These include sulfur, sand and clay to provide bulk, and starch and animal glue to hold the ingredients together. If the matches are to be self-igniting rather than safety matches, each head then gets an "eye"—a tiny drop of phosphorus. Last, the heads are dipped into formaldehyde to protect them from weather changes.

While still on the metal plates, the matches are dried. Then they are released and poured into a trough, which feeds them into boxes. The matches are made and packed in only one hour.

More matches are made in the United States than anywhere else in the world. In eight hours some factories make 200 million matches—enough to fill five freight cars.

PENCILS

The lead pencil is completely mis-named. First, the word "pencil" comes from the Latin wood *penicillum*, which means "little tail." It was the name for a small lettering brush used by the Romans to write on papyrus. Second, the lead in pencils is really graphite.

The history of the word "lead" goes back many years. The Greeks and Romans used small lead disks for making straight lines on papyrus to help them keep their lettering even. No one knows who first used a thin rod of lead as a writing tool, nor who first named it the lead pencil, but the name was used before the fourteenth century.

Since 1564, the "lead" in pencils has been graphite, a form of black carbon. Graphite was discovered accidentally. A tree blew down in Barrowdale in northern England, uncovering graphite. The graphite was so pure and hard that it could be used for writing. The English sawed it into square "leads" and put wooden shafts around it, thus making the ancestor of the pencil we use today.

Although graphite can be mined, most of the graphite used in today's pencils is made by heating either anthracite coal or petroleum coke,

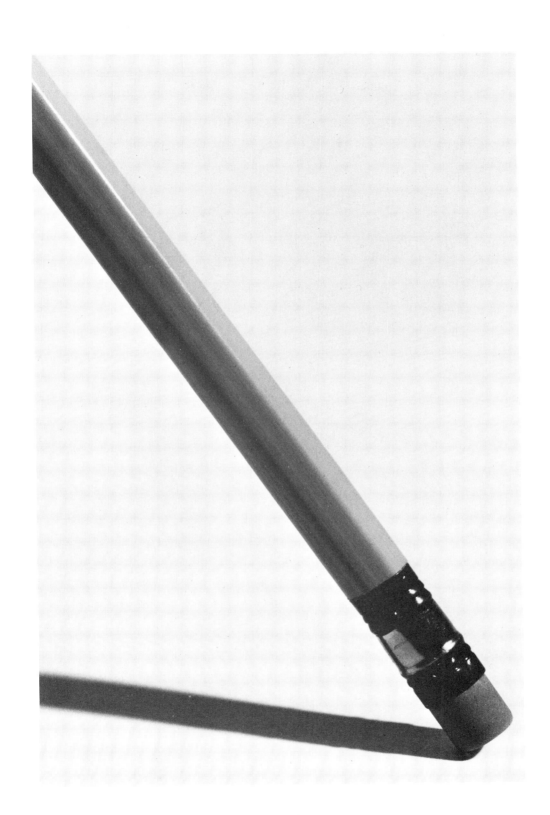

together with a small amount of ash, in an electric furnace. The graphite is then smashed into a very fine powder. This is mixed with water and with clay. The clay holds the graphite together. This wet mixture is poured into large round containers called drums. The drums rotate, grinding the mixture for weeks.

Then a machine presses the water out of the mixture. What's left is a graphite and clay "dough." The dough is put in an extruder, a machine which forces it out through an opening shaped like the finished lead. At this

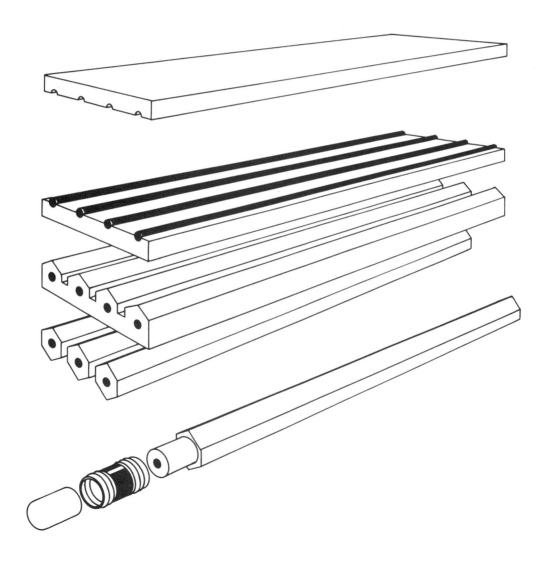

point, the lead looks like a long black shoestring. As it comes out, it is cut into 7-inch pieces.

These leads are too soft to use, however. To harden them, they are put into special pots which will withstand high heat and baked. The leads are now ready to be "sandwiched" between wooden slats.

Cedar wood is used for making pencils. It is a "soft" wood which whittles easily and has a straight grain. The cedar is cut into rectangles called slats. They are usually one pencil long, nine pencils wide, and half a pencil thick. (The slats shown in the diagram are only four pencils wide.) A machine cuts nine parallel grooves in each slat. Every groove is glued, and another machine puts a lead in each grove. Another slat, which is grooved and glued just like the first one, is placed on top. The two are dried under pressure. The result is a "lead sandwich."

Shaping machines carve the top half of each sandwich into pencil shapes. Then the machine flips the sandwich over and carves the other side, and the sandwich becomes nine pencils.

Each pencil then goes through a series of coating machines which give the pencils from four to eight coats of lacquer. Electrically heated stampers imprint each pencil with the brand name and pencil type.

The pencil is now ready for its eraser. There are five steps in putting an eraser on a pencil, and they are all done in one machine. A shoulder is cut on the pencil to hold the metal ring or ferrule. The ferrule is slipped on and clinched to the wood. The eraser is forced in, riveted, and washed. The pencil is now complete and ready to be used.

Altogether there are over 150 steps in the making of pencils, and the entire process takes three months.

One pencil can draw a line thirty-five miles long or write about 45,000 words. Two billion pencils are made in the United States every year. If all those pencils were laid end to end, they would reach more than 200,000 miles or about nine times around the earth at the equator.

BATS

As all ball players know, there is a big difference in bats. Every player in the major leagues has his bats made to order. During one season, he may go through as many as twelve of them. There are certain regulations for the way a bat must be made, but within those a professional player chooses what he wants. A player can decide on the length, the width, the taper or shape, and the weight. He can also choose the type of knob and the kind of grip he wants. These same designs are duplicated or scaled down in exact proportions and sold with the player's autograph for amateur use. The Mickey Mantle autograph has long been a big seller, and now Johnny Bench's bat is very popular.

Though some bats are made of aluminum, most are made from wood. The best wood for bats is northern white ash, because it is very strong. The trees are cut down and taken to a mill, where they are cut into lengths called bolts. These bolts are split and turned into "rounds" on lathes. The lathe holds the wood in place and rotates it while it is cut by special knives to whatever shape is needed.

These long, round pieces of wood are shipped by train to a timber yard. Here they are inspected, stacked, and seasoned.

14.

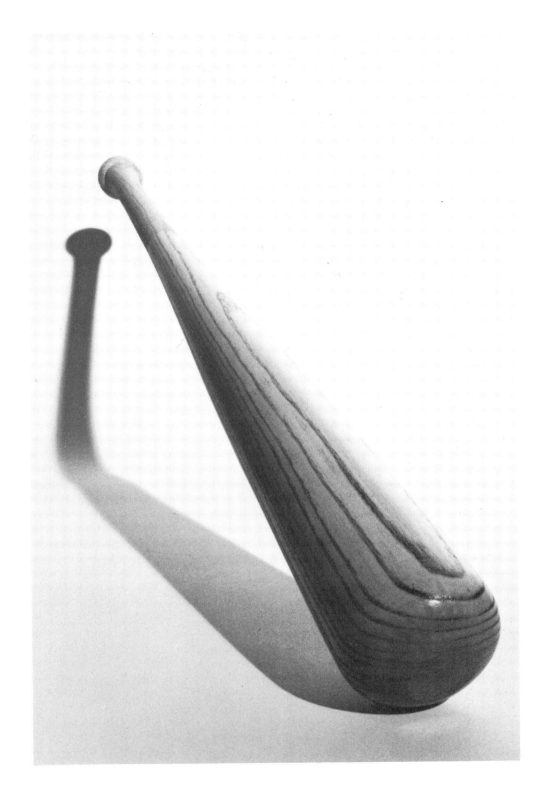

15.

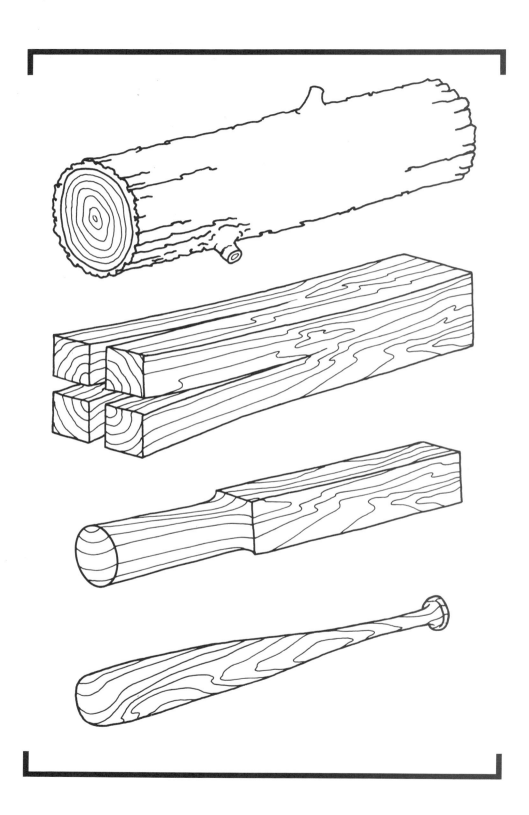

Seasoning is very important. Wood that has just been cut is called "green" wood. It is full of sap and gum that must be removed so the wood will be strong and keep its shape. When the wood is seasoned, it is left to dry in the open air. As the wood dries out, it shrinks and become stronger. This aging or seasoning takes one to two years.

When the wood, now called billets, has been seasoned, it is taken to a bat factory. The billets go into lathes and are made into rough bats, all of which are the same size. These roughouts are then inspected and weighed to determine what kind of bats they should be.

The roughout goes into another lathe. The lathe operator then checks the files to find out exactly how the player wants his bats made. He then sets the knives on the lathe. This takes great skill, for the knives must be set so that the bat to be made will be exactly the size, length, and taper ordered. After the knives are set, it takes only a few seconds to turn the roughout into the bat which has been ordered.

Next the bat is put into a sanding machine to smooth the wood. Then it is carefully inspected, and, if approved, it is given a clear lacquer finish. A machine burns the brand name on it. If the bat is for a professional player, his autograph is stamped on. For softball bats, which are narrower and shorter, tape is wound around the handle by machine.

The bats are now ready to be shipped and used by sandlot players and professional players across the country.*

---

* Bat shall be a small round stick not more than 2¾ inches in diameter at the thickest part, and not more than 42 inches in length.
The bat shall be:
one piece of solid wood,
or formed from a block of wood consisting of two or more pieces of wood bound together by an adhesive and in such a way that the grain direction of all pieces is essentially parallel to the length of the bat. Any such laminated bat shall contain only wood or adhesive, except for a clear finish.
No laminated bat shall be used in a professional game until the manufacturer has secured approval from the rules committee of his design and method of manufacturing. The bat handle for not more than eighteen inches from the end may be covered or treated with any material (including pine tar) to improve the grip.   ·
No colored bat may be used in a professional game unless approved by the said rules committee.
Cupped bats, an indentation in the end of the bat up to one inch in depth, is permitted and may be no wider than 2 inches and not less than 1 inch in diameter. The indentation must be curved with no foreign substance added.

Credit: *Official Rules of Baseball*, published by Sporting News.

TOOTHPICKS

Toothpicks are simply small pieces of wood that have been smoothed and polished. For centuries, Brazilians took pieces of wood, whittled them down to small splints, and used them to clean their teeth.

During the Civil War, an American named Charles E. Forster, who had been living in Brazil, brought the first wood toothpicks to the United States. Up until that time, quills and matches sharpened with a pocket knife had been used, and the rich sometimes carried gold toothpicks on watch chains. Convinced that toothpicks would be immediately popular, Forster, with another man, Charles C. Freeman, invented the first toothpick-making machines and opened, in Boston, the first toothpick factory.

However, people continued to use quills and matches. To drum up interest in wooden toothpicks, Forster and Freeman hired a well-dressed man and paid him to eat dinner in the best Boston restaurants. After each meal he would demand a wooden toothpick; when the restaurants said they didn't have any, he would create a scene, saying loudly that no restaurant could be a good one if it didn't carry toothpicks. It worked. Before long,

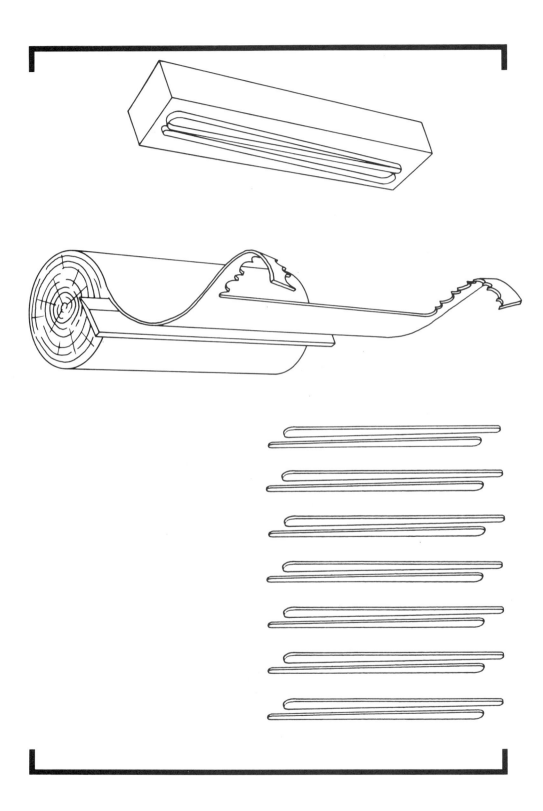

Bostonians were using wood toothpicks, and the idea spread to the rest of the country.

The making of toothpicks actually begins with the cutting down of white birch trees in the forest. White birch is used because of its color and because it has almost no taste.

The birches are cut into logs and delivered to the mill. Here the best-quality logs are selected for making flat toothpicks.

In the mill these logs are cut into 2-foot lengths, and the bark is mechanically peeled off each bolt. In the winter, these bolts of wood are thawed in large rooms filled with steam.

The logs are then put into a lathe. The lathe holds the bolt and turns it against a number of knives, which pull off long, thin layers of wood called veneer. These look like ribbons, and they are rolled up.

If flat toothpicks are being made, the rolls are fed into choppers. The veneer moves under a stamper which cuts out several thousand toothpicks each minute in just the same way that a cookie cutter stamps cookies out of dough.

If round toothpicks are being made, the veneer moves under a die which stamps out splints a little larger than toothpicks. These resemble wooden matchsticks except that their ends are slightly tapered. The splints are carried on a conveyor and are dropped into slotted boxes that line up the splints, end to end, in a series of long grooves. They are forced by air pressure into rounders, each of which contains a set of circular knives. The splint is forced between the rotating knives and is rounded into a toothpick. It then passes between a series of rotating, circular knives that point the ends.

The toothpicks, round or flat, are dried by putting them on a moving screen which passes through an oven. After they are dried, they are fed into a large revolving drum, where they are polished by tumbling against each other.

They are fed into machines which automatically count and package them in boxes and cases. It takes nine hours to convert a birch tree into toothpicks, and about 20 billion toothpicks are made each year in the United States.

PAPER

The word "paper" comes from the word "papyrus," a marsh plant. Ancient Egyptians used mats of woven papyrus as writing surfaces. Paper as we know it today was invented around two thousand years ago by a Chinese court official named Ts'ai Lun. He mixed hemp, and rags with water, mashed them into pulp, pressed out the liquid, and hung the "paper" to dry in the sun.

Most of our paper is made from wood pulp. Wood chips are cooked to make the pulp. The pulp is cleaned, dye is added in a beater, and the pulp is carried by water into a machine which converts it into paper, sucks out the water, and dries it.

But the making of paper begins in the woodlands. Trees are chopped down, cut into logs, and hauled to the wood yard of the paper company. They are put on a conveyor and fed into a giant revolving drum. Inside the drum, knives and powerful sprays of water or steam strip away the bark and clean off the wood.

The wood moves on to a chipper, which has a revolving steel disk with four or more sharp knives on it. It quickly cuts the logs into tiny chips.

To separate the cellulose fibers from the unwanted parts of the

wood (such as saps and resins), the wood chips are put into a pulp cooker, called a digester, which looks like a tall rocket ship and works like a pressure cooker. Chemicals are added, and the chips are cooked until they look like oatmeal. This cooking frees the cellulose fibers, known as the wood pulp. The pulp is then washed to remove the chemicals and other unwanted materials, and then it passes through a series of screens to remove unrefined pulp. If it is to be white paper, it is pumped to a tank where it is bleached.

The pulp is washed again, and then it is put into a beater. In this vat, the pulp passes between metal bars which cut the cellulose fibers into shorter lengths. The length of time that the pulp is beaten determines the uniformity and quality of the paper. During the beating, dyes may be added for colored paper, as well as chemicals to make the paper water-resistant.

Now the pulp goes into the Fourdrinier machine, which is a papermaking machine developed in France and introduced into England by Henry Fourdrinier in 1803. The Fourdrinier carries the pulp on a moving screen which allows the water to fall down away from the fiber. The screen shakes, causing the fibers to interlace, forming a stronger paper. The papermaking machine is 300 feet long; the pulp goes in the "wet end," and the paper comes out the "dry end" as one long, endless roll. The machine screens, drains, dries, and irons the paper.

The paper is pulled off the screen by rollers, sometimes at the rate of more than 3,000 feet per minute. The paper is then squeezed between heavy rollers and winds through a number of steam-heated rollers, which dry it.

If it is to have a smooth finish, the paper is ironed by heavy rollers called calenders. The paper comes out the "dry end" in large rolls which can be cut into sheets or slit into smaller rolls.

When paper was made by hand, a mark identifying the maker or the customer was often applied. It could be someone's initials, a royal crest, or a family seal. The mark was made while the paper was still wet, and came to be called a watermark.

Watermarks are still used, particularly on valuable documents to prevent counterfeiting and on stationery and typing paper for identification and advertising. They are applied to the still-wet flowing web of paper by a roller which presses a raised copper or rubber mark into the paper.

Approximately 60 million tons of paper a year are manufactured in the United States. To manufacture this amount of paper, the paper companies in this country use about fifteen million tons a year of recycled paper, most of which is bought from waste paper dealers. Recycled paper is paper that has been used, reclaimed, de-inked, washed, bleached and beaten into pulp again, to be mixed with virgin pulp. U.S. consumption is estimated at 600 pounds per person. Of this, 225 pounds of paper go into reading and writing material and about 35 pounds go into tissue, mostly facial and toilet tissue. The remaining 340 pounds is used mostly for packaging and industrial purposes.

PAPER BAGS

The earliest paper bag was a sheet of paper formed by hand into a cone. The point was folded over so that it would hold together, but the customer often found it difficult to get home without its collapsing and spilling out his purchases.

In 1851, a man named Francis Wolle invented the first paper-bag machine. As a teen-ager he had worked in his father's store and every evening had had the job of cutting and pasting paper bags for the next day's sales. He had never liked doing it, and seventeen years later, he invented a machine which could automatically make paper bags. He put a roll of paper on one end of the machine and started it turning. The paper was pulled onto a level area, and a knife cut it into pieces all the same size. The sheets were picked up mechanically, the edges were folded, and the paste was applied.

Today's paper-bag machines work in a similar way, only now there are different machines for every size and shape of bag. The first machine made a flat bag—that is, one with the sides simply glued together, like a manila envelope. Flat bags are still made and used for all sorts of purposes, but the most common bag today is the square bag used in supermarkets. It is sometimes called an "automatic" bag, because a grocer or clerk can

27.

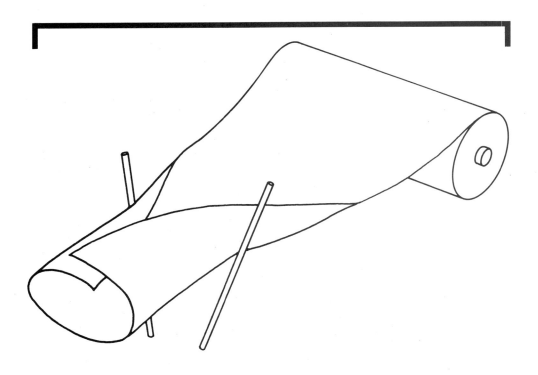

put his thumb in the notch opening and snap the bag quickly, and it will open and stand by itself.

To make the square bag, a huge roll of paper is mounted on the back of the bag machine. The paper is pulled through steel rollers. A small press prints the size of the bag on it, and paste is put on close to one edge.

The sheet of paper is drawn over forming plates which fold the paper into one long, continuous flat tube and also press in the sides, or tucks, of the bag. As the tube moves along, the edges are ironed into sharp creases and the seam is formed from the pasted edge.

Knives cut the thumb notch as well as a pair of slits into the tube. These slits will later form the bottom. Two blades also score the tube—that is, they cut across it but not through it in such a way that the bottom of the bag is formed.

Metal "fingers" pick up the tube and pull it onto a revolving drum. The tube is cut to the length of the bag and is opened. The slits in the tube are folded to form the inside bottom of the bag.

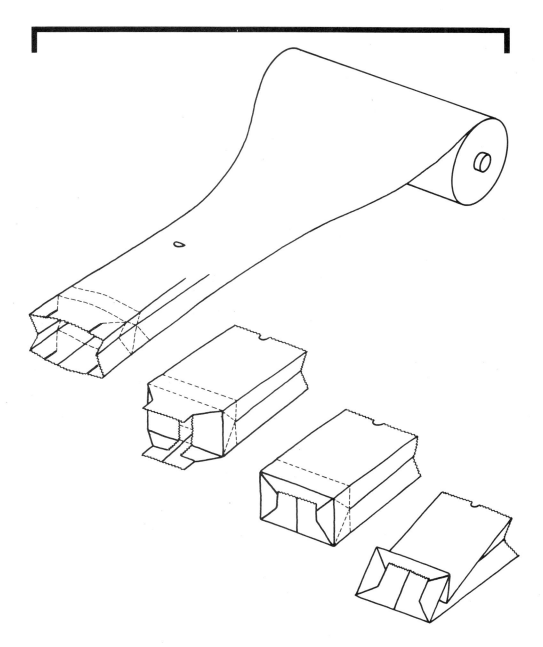

The bag is carried on the drum past a roll which puts paste on the bottom. The bottom edges of the bag are tucked in. Half-circle "wings" on either side of the drum close and press the pasted bottom flaps together. Pressing rollers iron the bag and out it comes to the stacking table, where it is quickly followed by the next bag and the next. In one minute, 450 paper bags can be made.

CLAY

BRICKS

Bricks are the oldest man-made building material. They have been used for at least six thousand years; the Babylonians and the Egyptians first used them for religious buildings as well as homes. When the Israelites were held captive in Egypt, one of their tasks was to make bricks.

In those days, bricks were made by crushing clay and mixing it with water. Workmen did this by stomping barefoot on the clay. Straw was added to the mixture to help hold it together and to make the bricks stronger. The mixture was formed by hand into different shapes, and then these bricks were allowed to dry in the sun.

Today's bricks are still made from clay, but they are baked at high temperatures in ovens called kilns. This firing of the bricks strengthens them and fuses the clay together, so straw is no longer needed.

To make bricks, chunks of clay are shipped directly to the factory from the mines. The chunks are crushed and ground by huge revolving wheels weighing from 4 to 8 tons each. These wheels revolve in a large round pan. To break the clay into smaller pieces, it is passed through vibrating screens to a large mixer that has revolving blades. Water is added, and it is mixed

33.

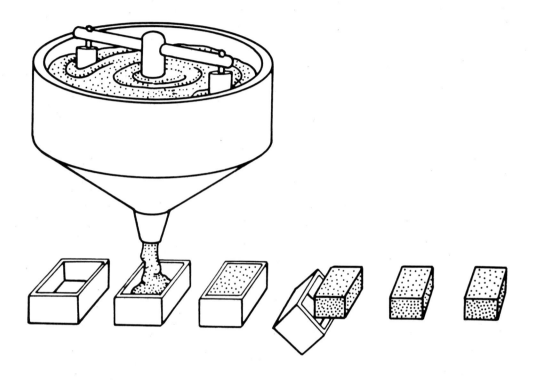

with the clay until it is smooth and ready to be molded. This is called pugging.

There are three ways of molding bricks: the soft-mud process, the stiff-mud process, and the dry-mud process. The difference is the amount of water in the clay.

The oldest way is the soft-mud process. Clay is put into molds, which are coated with sand or water to keep the bricks from sticking. This can be done by hand or by a press that molds several bricks at a time.

The newest and most frequently used method is the stiff-mud process. The clay contains less water than that used in the soft-mud process. In the stiff-mud process, the clay is put into a machine that takes out the air, making the clay stronger. Then it moves into an extruder that

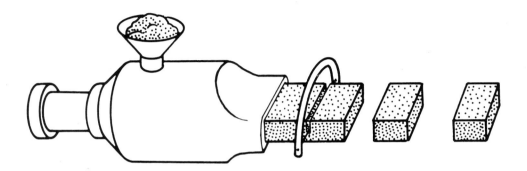

forces the clay out through an opening shaped like brick. As it comes out, steel wires automatically cut it into separate bricks. Imperfect bricks are put back in the mixer.

Bricks can also be made by the dry-press process. Relatively dry clay is mechanically pressed into molds.

Whether bricks are molded or extruded, water must be removed from them, so they are dried in kilns for a day or two.

The bricks are then fired—that is, baked at high temperatures. Depending on the type of kiln, this can take from 40 to 150 hours. The most common method is to place the bricks on special cars that move through a tunnel-shaped oven. The temperature is gradually raised from 400° to 2,400° Fahrenheit. The bricks are then slowly cooled to avoid cracking.

When cooled, the bricks are sorted by size, graded for quality, and then loaded into railroad cars or trucks.

Brick buildings can last hundreds of years. Bricks are a good building material because they are inexpensive, can support great weight, absorb little water, and can withstand both heat and cold. Moreover, their beautiful color and texture are often preferred to other materials.

CHINA CUPS

In prehistoric times clay cups were made by hand. Designs were made on them by using a pointed stick or a finger. They were then placed in an open wood fire to harden.

Porcelain cups—that is, cups made of clay and china stone—were first made in China about a thousand years ago, and thus the use of the word "china" for the cups we use today. The basic methods and materials have changed very little.

Clay, flint, feldspar, and china stone are ground and crushed until they become fine powders. In a large tank, the powders are blended with water into a smooth creamy liquid. This clay mix moves through fine screens that remove the impurities. It is then pumped into a magnetized trough that removes any iron. The refined clay mixture is pumped into a filter press, where the water drains through nylon cloths, leaving a slab of clay weighing about 100 pounds.

These slabs are kneaded, or "pugged," until smooth in a mill. The clay is now forced out of the mill in an endless roll and then cut into pieces for the potters.

To make a cup, the potter takes a ball of clay and throws it into a

37.

plaster-of-Paris mold, which forms the outside of the cup. The mold rotates, and a metal form is lowered to shape the inside of the cup.

Cup handles are made in molds or from clay strips squeezed through a machine and then shaped by hand. The potter takes a bit of clay mixed with water and presses the handle on.

When dry, the cups are carefully stacked on special platforms and fired in a kiln for about fifty hours.

These cups, which are now hard, are dipped by hand into a liquid glaze. The glaze is actually a finely ground mixture of frit, clay and flint in water. If the cups are to be colored, pigment is added. After being glazed, the cups must dry. The water in the mixture evaporates, leaving a thin film of shiny glaze.

The cups are again placed in a kiln and fired for about thirty hours. This fuses or "marries" the glaze to the cup.

China cups are usually decorated after they are glazed, and then they are fired again to fuse the design into the glaze.

One of the most popular ways of decorating cups and other pottery is by using multicolored transfers with adhesive backing. These are like decals.

At the china factory the sheets of transfers are dipped in water and the transfers are lifted away from the backing and carefully put around the cup.

Another type of transfer, the plain print, can also be used. This is a tissue-paper transfer that is put face down on the cup. It is firmly rubbed first with a cloth and then with a brush. Then the paper is washed off, leaving the pattern on the cup. Some very fine china, however, is painted by hand.

After being fired again, the cups are ready to be shipped. To avoid breakage, the cups are carefully packed by hand.

RUBBER

The Indians in Central and South America were bouncing rubber balls long before Columbus landed. And over twenty-six centuries ago, Egyptians and Ethiopians made wall drawings showing people bouncing balls.

For a long time no one knew what else to do with the sticky, milky liquid that oozed out of certain trees. Spanish and Portuguese explorers tried to use it to waterproof their clothes. But when the sun came out, it melted all over them instead.

In 1770, Dr. Joseph Priestley, an English scientist, learned that he could use this sticky substance to erase pencil marks. He called it "rubber" because it could rub marks from a page.

In 1839 an American, Charles Goodyear, mixed sulfur with rubber as part of his continuing experiments to make natural rubber less sticky and more dry. He accidentally dropped the mixture on a hot stove. It quickly fried and turned hard and tough. This process is known as vulcanization, after Vulcan, the Roman god of fire. After vulcanization, rubber will not turn sticky and thus has many more uses. Goodyear was a better inventor than businessman, however, because patents on his discovery were never properly completed and he died in debt. The rubber company with his name chose it to honor his role in the history of the rubber industry.

A crude process for making synthetic rubber was discovered in 1867, but it wasn't widely used or developed because natural rubber was so abundant and so cheap. When World War II closed U.S. access to the Far East, our major suppliers of natural rubber, the synthetic rubber industry sprang up out of necessity in the United States. Today, synthetic rubber is made from petroleum, alcohol, coal tar, turpentine, tar, and natural gas through complicated chemical processes similar to those used to make plastics. (Plastics are the subject of the next part of this book.) Some synthetic rubber duplicates exactly the molecular structure of natural rubber. It accounts for more than 75 percent of the rubber we use.

There are dozens of kinds of synthetic rubber. Each of them has special properties that makes it suitable for special uses. Some have properties that natural rubber doesn't—some are resistant to grease, oils, and solvents, others to air, and others to extreme cold.

Natural rubber, however, has better elasticity. Although synthetic rubber is used in making car tires, natural rubber is often preferred for large truck, bus, and airplane tires because of its heat-resisting properties.

Most natural rubber comes from Indonesia, the Philippines, Brazil, and Guatemala. It is collected much the same way as it was when the Indians did it. A thin, slanted strip of bark is cut from the rubber tree, and a liquid called latex runs down the slanted cut and drips into a cup.

The latex is sent to a factory, where the water is taken out of it. This is done by adding formic acid to the latex to make it thick and then putting it between rollers to squeeze the water out. The rubber is then chopped into small pieces and dried like a load of clothes in a dryer.

After it is dried, the rubber is pressed into bales and shipped. Then, depending on what is going to be made from it, the rubber may be ground, dissolved, and mixed with chemicals and pigments for coloring. It may be shaped, molded, or used as a coating for another material.

TIRES

There are nearly one thousand kinds and sizes of tires. We see them on cars every day, but there are also tires for many other kinds of machinery. We have become a "nation of wheels," and it would be hard for us to survive without tires.

Tires are made from synthetic rubber mixed with chemicals, fabric, and metal. Each material is prepared separately and then combined by a tire manufacturer. Different manufacturers use different methods, but basically this is how it's done:

A number of different combinations of rubber and chemical mixtures are used to make tires. The rubber for the treads differs from the rubber used for the sidewalls or in the coating of fabric inside the tires. The tread rubber must be tough and resistant to wear. The sidewall rubber must be strong enough to support the vehicle and hold the tire together. The rubber coating for the fabric must be resilient. Different chemicals are used for each, but they are all made the same way.

Rubber is mixed with carbon black and chemicals in a huge

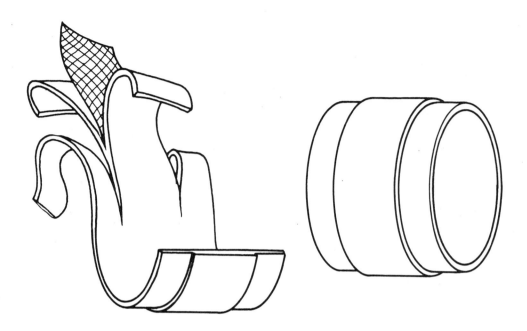

enclosed mixer, a machine with sharp, rotating blades. Then the rubber is fed into a milling machine that has metal rollers. The rubber is forced through these rollers, mixing the chemicals more thoroughly. This causes the rubber to get hot. To keep it from burning, cold water flows through the hollow rollers.

Next, the rubber compounds for the treads and sidewalls are both fed into the same machine. It pushes them out as one long strip. The tread for the tire is in the center, and the sidewalls are on either side. Then this long strip of tread and sidewall rubber is coated with an adhesive. This will be used later to hold the tread and the sidewalls to the body of the tire. Once the strip has cooled and shrunk, it is cut to exact size.

Fabric gives the tire its strength. Polyester, nylon, and rayon are used in making car tires. Rolls of the fabric are specially woven and sent to the tire factory. A machine coats the fabric with rubber. Then it goes to a cutter and is cut into strips called plies. A ply is a layer of fabric. A two-ply tire is made with two layers of fabric.

While the rubber is being prepared, the "beads" are being made. The bead holds the tire on the rim. It is made up of strands of steel wire

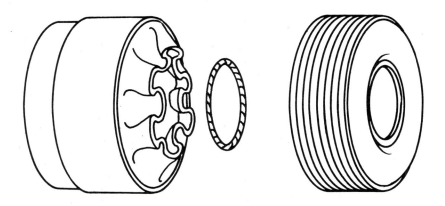

which are braided together. They are coated with rubber and then wound into a circle.

The plies, the tread, the beads, and the sidewalls are put together on a machine which is like an inflatable drum. The drum is really a frame for the tire. It is inflated into a circle the size of the tire being made. The rubberized strips of fabric—the plies—are put on the rotating drum. Then the beads are attached and stitched into place by machine. Finally the tread and sidewall rubbers, which have been coated with adhesive, are added in one piece.

When the drum is collapsed, the tire comes off, looking like a barrel with open ends. The tire is ready for the curing room, where it is placed in a mold and shaped. Heat and pressure are applied to make it strong and hard. (This is vulcanization, discussed in the previous chapter.) The mold gives the tread and sidewalls their designs.

Depending on the size of the tire, the curing can take from fifteen minutes to eighteen hours. Finally the tire is complete and "ready to roll."

Rubber companies make tires from 3 inches in diameter for toys to 12 feet in diameter for 200-ton dump trucks.

RUBBER BANDS

In many ways, rubber bands have replaced string or ribbons for holding things together. We use them to hold tops on boxes, to keep papers together, and to pull hair back into ponytails. We could live without them, but they are one of the simple devices that make our lives easier.

Rubber bands are made by mixing raw rubber together with pigments to color it and chemicals to strengthen it. This mixture, compound rubber, is fed through a funnel into an extruder. This works something like a meat grinder. The rubber is fed in at one end, ground and kneaded in the machine, and forced out an opening which forms it into a long tube. Then it is cut into 7-to-10-foot lengths.

Rubber at this stage is soft and sticky and will not keep its shape. It needs to be cooked, or vulcanized, to make it strong and hard. So, the tubes of rubber are put on long metal rods and hung from racks inside a large boiler. For twenty minutes, the tubes are cooked by steam heat.

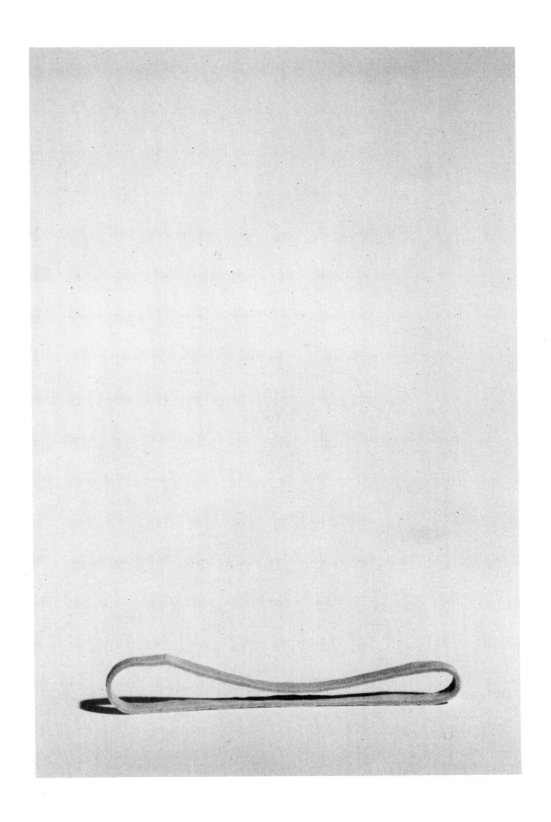

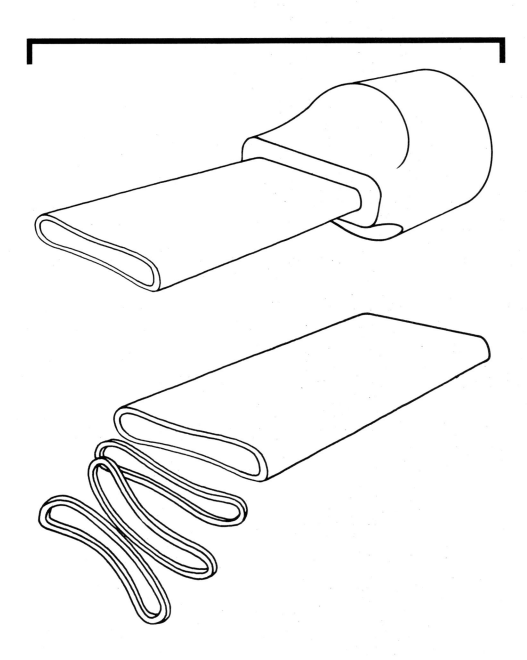

The tubes are taken out and are automatically cut into rubber bands by a series of blades mounted on a wheel. The rubber bands can be from 1/16 inch to 1 inch wide.

They are now ready to be used for everything from holding dental braces in place to making slingshots.

BALLOONS

Whether balloons are round, long, oval, twisted, or shaped like animal heads, they are all made on aluminum or porcelain forms which are the shape of the balloon *before* it is inflated. The forms are hung upside down on metal strips in groups of eight to twenty and move automatically from one tank to another.

The first stop for the forms is the coagulant tank. They are dipped in a mixture of calcium, talc, and alcohol. This mixture will make the natural latex, or liquid rubber, stick to the forms.

The forms travel next to a colorful part of the factory, the dipping tanks. Each tank contains a different-colored latex—red, green, blue, pink, yellow, orange, or white. Each set of forms is lowered into the appropriate tank. When the forms are lifted out of the tank they are covered with latex. They move between two spiral-shaped power brushes that roll the liquid rubber into a ridge or ring for the neck of the balloon.

The latex-covered forms then go through hot-water tanks to be washed and from there into a drying oven. After they dry, they are dipped in another tank filled with soap and water to keep them from sticking together when they are packaged.

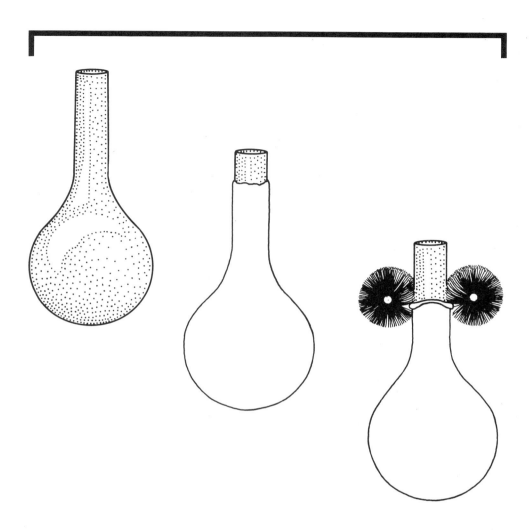

The balloons are taken off the forms by hand, with some help from short blasts of air from air jets. They are dried again in a tumble dryer.

Many balloons have printing on them, from animal faces to advertisements. If a balloon is to be printed, it is blown up by a jet of air and put on a spindle to keep the air from escaping. The design is quickly stamped on. The balloon is removed from the spindle and is allowed to deflate. A conveyor belt takes the balloon through an oven to shrink it back to size.

The making of a balloon takes about fifty minutes.

According to the *Guinness Book of World Records*, Jane Dorst released a balloon in Atherton, California in May, 1972. Two and one-half weeks later it was found in Pietermaritzburg, South Africa, nine thousand miles away.

SNEAKERS

Indians in Brazil were making their own form of sneakers three hundred years ago. They simple plunged their bare feet into latex, the milky fluid from the rubber tree. These "shoes," however, did not last long. And when they wore out, they had to be peeled off the feet.

Today's sneakers (also called tennis shoes or gym shoes) contain as many as thirty-five pieces of rubber and fabric that are stitched and glued together.

The soles of sneakers are made of rubber. Raw rubber and chemicals are heated and blended in a mixer for about eight minutes. Then the rubber flows onto a set of rollers that press and cut it into strips. The strips are ground up in a machine, reheated, and melted. The liquid rubber is put into molds the shape of a shoe sole and allowed to harden. This takes only eighty-five seconds. The top and sides of the sole are then coated with liquid rubber, which acts like glue.

The "upper," or the top of the shoe, is usually made of canvas, although nylon and leather are becoming more popular. Heavy metal dies, which resemble giant cookie cutters, are used to cut any one of these fabrics.

55.

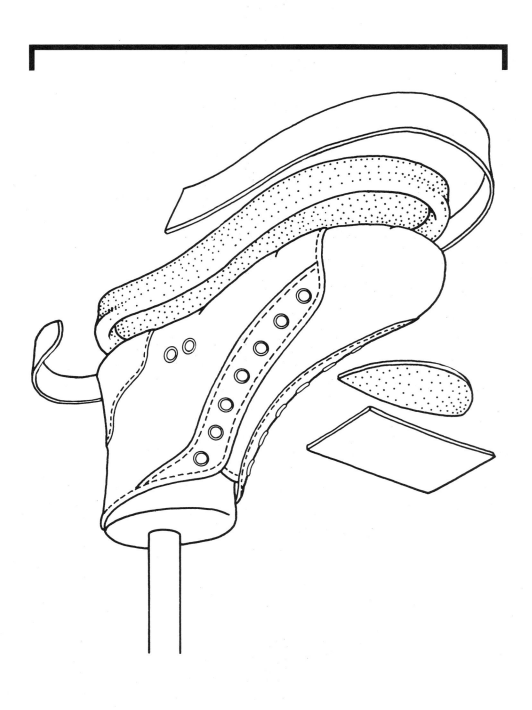

The dies are placed on the layers of fabric. A heavy beam comes down and pounds the dies so they cut through the fabric. One die can cut through twelve to twenty-four layers of fabric at a time.

The top of the shoe is really many pieces of fabric stitched together on an assembly line. One person sews the back seam together, another sews the binding around the edge, another punches eyelets on a machine, and so on. When the pieces arrive at the end of the assembly line, they look like shoes without soles.

The uppers are now ready to come together with the soles. This is done on "lasts," which are aluminum, cast-iron, or plastic feet. As the fake foot begins moving on the assembly line, one part of the shoe is put on after another. First the upper and the insole (the inner lining) go on. Next the bottom edges of the upper are folded around the last onto the insole, which is coated with glue. The bottom of the shoe is then dipped in latex so that the rubber sole can be bound to the shoe. At each stop something else is added: a rubber toe guard, the rubber strip around the shoe, the label. During the process the shoe is squeezed in clamps several times and pressed with rollers to keep the pieces stuck together.

Still on their lasts, the shoes are placed on racks with wheels and rolled into a vulcanizer. This huge pressure cooker cures, or cooks, over nine hundred pairs of sneakers for an hour.

When ready, the shoes are removed from the lasts and inspected. They will be shipped to stores in twenty-three countries. Americans buy 215 million pairs each year.

PLASTIC

Plastics are man-made materials. They are made from chemicals which come from such raw materials as coal, petroleum, limestone, and salt. There are more than forty "families" of plastics, and within each family there are dozens of formulas for particular uses. Plastics can be hard or soft, transparent or opaque. They can be molded into many shapes—dishes, combs, telephones; they can be used to make squeeze bottles, garbage cans, inflatable toys; and they can be used to make transparent wrappings for fruits and meats, to make shower curtains, to make eyeglass lenses. They have thousands of uses.

One of the most common plastics is polyethylene. Squeeze bottles, dry cleaners' bags, ice-cube trays, food-storage containers, and freezer bags are made from it.

Polyethylene is made from ethylene, a gas derived from petroleum. The gas is stored in cylindrical containers. It is piped from these storage units into the top of a reactor chamber. This looks like a tall, thin tower. The inside of the reactor looks like a series of platforms separated by screens. The screens have been coated with a catalyst, usually a metallic salt. A catalyst is an ingredient that causes changes in other ingredients without being changed itself.

Intense pressure and heat from a compressor forces the gas down through the screens. As it passes through the screens, it reacts to the catalyst and gradually changes from a gas into a thick liquid plastic. This liquid plastic is pushed out of the bottom of the reactor through an extruder. A screw in the extruder pushes the plastic out through a series of small holes. As it comes out, the plastic looks like long strands of spaghetti. Rotating knives cut the strands into pellets.

Basically what has happened is a chemical reaction called polymerization. All matter is made up of tiny invisible particles called molecules. In polymerization, small molecules called monomers join together to form large molecules called polymers. In other words, the molecules in ethylene join together to form a new material, polyethylene. It is like a chain. Each molecule of ethylene is a single link; when a number of them are put together, they form a chain of polyethylene.

The polyethylene in pellet form is the raw material used to make many plastic products.

STRAWS

Drinking straws are made from plastic or paper. Most straws today are plastic and are made by a machine called an extruder.

Tiny plastic pellets are dropped into a funnel or giant hopper on top of the extruder. The pellets are moved along the barrel of the machine by a rotating screw. Heaters inside the barrel melt the plastic, which is then forced out through a ring-shaped opening.

The melted plastic comes out of the extruder as one long, endless straw. The plastic must be cooled, so as it comes out, it is "pulled" by rollers on a conveyor through a long water bath. After it is cool, the long tube passes under a rotating knife that cuts it into straws of equal length.

For colored straws, small colored pellets are added to the plastic pellets in the hopper, at the beginning of the process. Once inside the extruder, they melt and evenly color the plastic. However, for striped straws, colored liquid plastic is fed into two holes in the machine, just before the straw is released from the extruder but after it is shaped, so the straw is released with two stripes winding down its length.

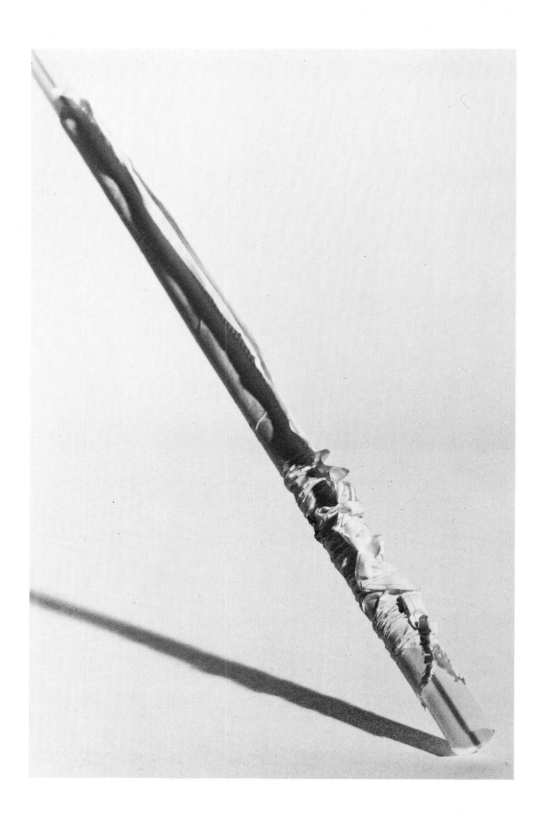

62.

The straws are then automatically packaged, either in boxes or, for restaurant straws, in individual paper wrappings.

Seventy-five straws can be made in a minute.

To make flexible plastic straws, a machine molds the "bend" in the straws after they have been extruded. A metal rod is put inside the straw, and the straw is rotated as it is pushed against a mold. The mold puts grooves into the part of the straw that will eventually bend. Both ends of the straw are then held firmly and pushed toward each other. The pressure forces the straw to "give" where it has been grooved, creating accordionlike pleats that allow the straw to bend.

ADHESIVE BANDAGES

Adhesive bandages* have been around for a long time in one form or another. The use of plasters or pastes to cover wounds can be traced from the beginnings of recorded medical history, as early as 3000 B.C. in Egypt. Olive oil, beeswax, and India rubber have all been used as protective coverings. Improvements on these coverings over the years have been to make bandages that are waterproof, that allow air to reach the wound, that stick faster, and that can be removed cleanly.

All adhesive bandages are made of a strip of material coated with adhesive and a gauze pad to actually cover the wound that is to be protected. There are different materials used for the strips that make up the body of the bandages; some are made of cloth, some of paper, some of tinted vinyl, and some of a clear vinyl film, which, being the most transparent material, is the least noticeable.

The body of a typical adhesive bandage, be it cloth, paper, or vinyl, starts as a big sheet which is coated with adhesive and then backed

---

* Modern adhesive bandages are often referred to as Band-Aids which is in fact a name that is a trademark for Johnson & Johnson Band-Aid brand adhesive bandages.

65.

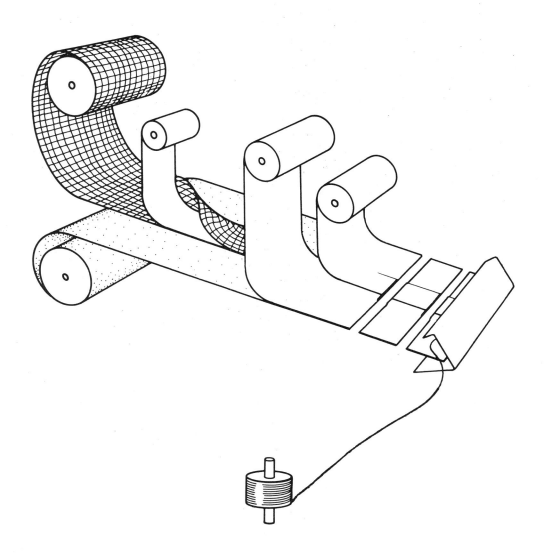

with a slick protective paper. The sheet is then cut into long strips of the desired width and wound into rolls. The backing paper keeps each strip from sticking to itself.

Then the vinyl (or other material) rolls—along with rolls of gauze, rayon filler, and paper for the covering tabs—are loaded into an assembly machine that puts the adhesive bandage together, cuts it, and wraps it. As the strip of vinyl is unrolled into the machine, air vent holes may be punched into it. Meanwhile, the rayon filler—which adds protective padding and absorbency to the bandage—is covered with a layer of gauze, cut, and placed along the center of the strip of vinyl. Next, the paper for the tabs is unrolled over the long strip of bandages, and blades cut the strip into individual bandages. In Band-Aids, a red string is laid along the length of each bandage and the whole thing is wrapped in glassine paper and boxed. The string later serves to help open the individual wrapper. When an end is torn off, the string can be pulled along the length of the wrapper to open it and free the bandage.

Adhesive bandages are then put in a sterilizer, an enclosure big enough to hold a small car. The sterilizer works like a pressure cooker, applying steam to the bandages and destroying any bacteria that might be present.

Every time a group of boxes of adhesive bandages is sterilized, one of the containers near the center of the group is actually inoculated with bacteria. After sterilization, this sample container is taken to a laboratory for testing to be sure all the bacteria have been killed. If they have been killed, it provides assurance that any other bacteria that may have found their way into the other boxes of bandages will also have been destroyed.

All the bandages in the batch are stored until the testing is completed. Then they are shipped to stores. About 4 billion adhesive bandages are sold every year.

TOOTHBRUSHES

Throughout recorded history, people have cleaned their teeth. The Greeks and Romans did it. And before the time of Christ, the people of India used twigs or leaves to clean their teeth. Other cultures used twigs which they frayed at one end, creating primitive brushes with handles. There are records of the Chinese in the fifteenth century using handles with brushes inserted in them. And in the sixteenth century, the English nobility used silver toothbrushes.

Today's toothbrushes are not that elegant. The handles are made from plastic, and the brushes are made of natural boar bristle or man-made nylon bristle. Toothbrushes are put together entirely by machine.

Small plastic pellets are fed into an injection molding machine. It heats the plastic until it melts. This liquid is forced by a plunger or by a rotating screw directly into closed molds which are in the shape of the toothbrush handle. Pressure is applied to the molds, which are clamped in place. After they are cooled, the clamps are released,

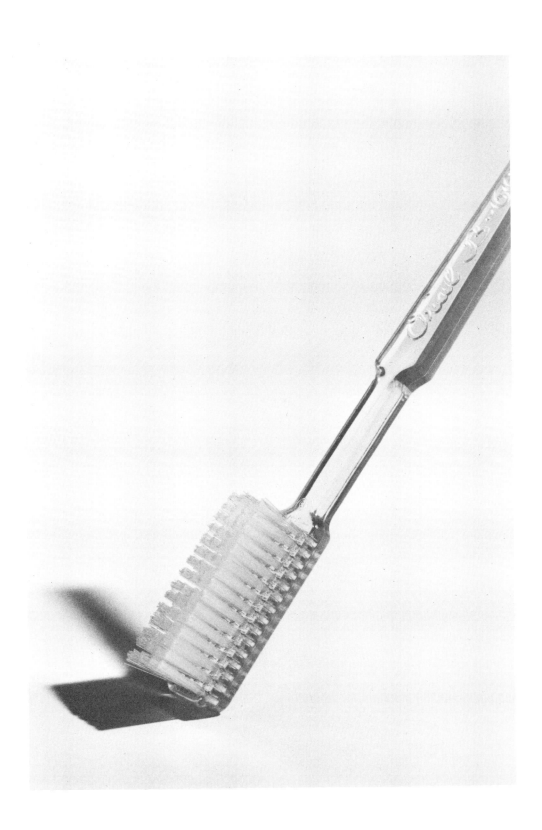

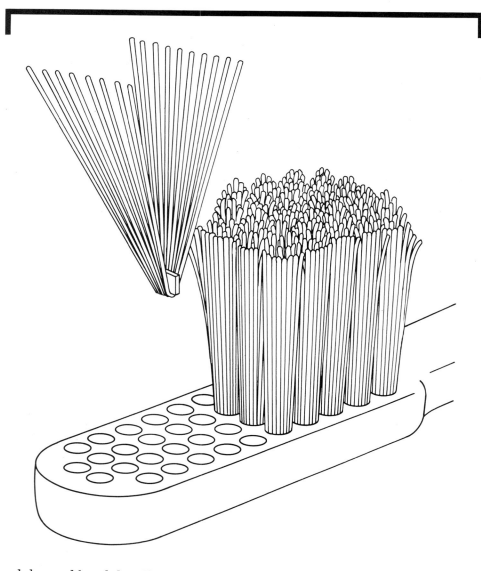

and the toothbrush handles are ejected from the molds by tiny pins. They are completely formed even to the point of having holes, or cores, which will hold the bristles.

The handles are then taken to the toothbrush filling machine. The right amount of bristle is inserted into each core, and the machine automatically forces the bristle into the core with very small metal staples which hold them in.

An automatic trimming device cuts the bristles to the desired angle. The toothbrush is now ready to be packed, shipped, bought, and used.

RECORDS

The first words to be recorded on a phonograph record were "Mary had a little lamb." This was in 1877, and the verse was recited by Thomas Alva Edison, the inventor of the first phonograph. The "record" was a cylinder covered with tinfoil. This tinfoil record wore out too fast, so the tinfoil was replaced by a new material, wax-coated paper, invented by Alexander Graham Bell and Charles Tainter in 1885.

Two years later, in 1887, Emile Berliner was able to record onto a disk. The machine which played this disk was named a "gramophone." Record players in England are still referred to as gramophones.

Until 1950, records were made out of shellac. These records were known as 78's because they were recorded and played at 78 revolutions per minute. In 1948, a new type of recording was developed, the long-playing record, or LP. Instead of getting only three minutes of music on a 12-inch record, it was now possible to have thirty minutes on each side. The old shellac records broke easily. So when the recording industry introduced the new 33⅓'s and 45's, they also improved their product by using the new vinyl plastics, which were unbreakable.

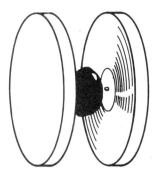  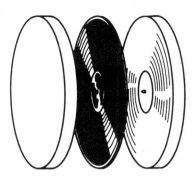

Today, music is first recorded on a tape recorder. By electrically connecting the tape record to a record cutter, the sound vibrations are transferred to the stylus or cutting head of the record cutter. The stylus cuts grooves into a lacquer-coated metal disc on the record cutter. These grooves are really the moving track of vibrations. The grooves start at the outside of the record and circle in to the center. There can be as many as 2,400 grooves on one side of a 10-inch LP.

This lacquer disc is dusted with gold and plated electrically in a copper or nickel bath. From this master record, metal molds are made that can be used to make thousands of records. The master is then put away so that an original copy of the record will always be available.

The molds are placed in a record press, which resembles a waffle iron. A gob of heated plastic, called a vinyl biscuit, is then inserted into the press and adhesive-backed printed labels are placed in the center of the molds. The molds are then forced together at a very high temperature. Then the molded record is removed, and its edges are trimmed off.

The records are then inspected, packed, and shipped to the stores, the juke-box companies, and the disk jockeys who will play them on the radio.

GLASS

Glass was first formed by volcanic fires in rocklike chunks called obsidian and made of three natural ingredients—sand, soda, and lime. Today we use over fifty chemical elements to make glass, but the three basic ingredients are still the same, with every mixture approximately 70 percent sand. There are different formulas for different kinds of glass, depending on what it will be used to make.

Sand, soda, and lime, which all resemble dry powders, are stored in large silos. Careful measurements of each are taken from each silo and then combined with cullet in a power mixer. Cullet is cleaned and crushed scrap glass. In bottle making, the cullet is recycled, crushed bottles.

The mix travels in carts on a broad conveyor to the furnace where it is melted at 2,700° Fahrenheit. If you could see inside the furnace—and some companies use special closed-circuit television cameras so they can—you would see a fiery boiling lake of melted glass.

The molten glass then moves into a slightly cooler part of the furnace and is ready to be formed into many products.

Depending on what it is to be used for, glass is formed in many ways. It can be blown by hand or by machine, or it can be pressed into molds, as described in the chapter about the bottle. It can be drawn out of the furnace as a wide sheet of glass, or it can be floated on a bath of melted tin, as described in the chapter on flat glass.

Immediately after being formed, glass products go through a process called annealing. They are reheated in ovens called lehrs and then slowly cooled according to specific time and temperature schedules. This heating and cooling changes the composition of the glass, removing internal strains and stress that might have been caused by unequal heating of the glass while it was being made. If not annealed, the glass could break or shatter very easily.

Glass can also be tempered after it is formed. It is heated until almost soft, then chilled by blasts of air or by plunging it into a cold liquid. Tempering changes the internal structure of the glass, making it much stronger than regular glass.

Thousands of glass products are made every year, ranging from simple window panes and soda bottles to lenses for giant telescopes.

For many years flat glass was made by glassblowers. The glassblower would stick an iron blowpipe into melted glass, gather a gob of it on the blowpipe, and blow it into a pit which was shaped like a cylinder. Blowing forced the glass up against the sides of the pit, shaping it into a cylinder that was about 4 to 6 feet long and 12 to 15 inches in diameter. The ends were removed from the cylinder and it was split down one side and put into an oven. The heat from the oven caused the cylinder to open and flatten out. But the craftsmen could not make the glass completely flat or free of marks.

Over the years many other methods have been used to try to make glass that is perfectly flat, has no flaws, and has a brilliant finish.

Today there are three types of flat glass: sheet, plate, and float glass. Sheet glass is used in regular windows; plate and float glass are used where very clear glass is needed, such as for mirrors, windshields, and sliding doors.

To make plate glass, hot, molten glass flows directly from the furnace between rollers that form it into a wide sheet. The glass goes through an annealing oven, where it is heated and slowly cooled to eliminate strains and stresses. The glass then moves between grinders and polishers, large flat discs which remove imperfections and give it a smooth finish. Then it is cut to size. Sheet glass is made in basically the same way, but it is not ground or polished.

The newest, and best, way of making flat glass is the "float process," which has been used for about twenty years.

Hot melted glass comes out of the furnace and is fed directly onto a perfectly flat surface of melted tin inside the "float bath." Gravity causes the glass to float on top of the tin, and heat from the melted tin and from heaters causes the glass to spread evenly.

As the glass moves across the tin, additional heaters give it a brilliant finish. Then it moves into a cooler section of the float bath where it becomes hard enough to be moved without being scratched by the rollers that carry it into the annealing oven. Here it is reheated and slowly cooled so that it will not shatter or crack. The glass leaves the oven in a long, flat ribbon and continues to the cutting area for cutting and packaging.

MIRRORS

A mirror is really glass with a metallic coating on one side. Primitive mirrors have existed for thousands of years. The first mirror was made by polishing obsidian, the natural glass formed by volcanoes.

The ancestor of today's mirror was first made in the fourteenth century in Venice. There, a craftsman discovered that by heating a flat piece of glass and coating one side with tin and mercury, he created a surface that reflected his own image better than anything else used at that time.

Today's mirrors are still made by coating one side of sheets of glass with chemicals, but the chemicals used have changed.

In the mirror factory, the glass rides on conveyors under a number of nozzles and sprays and brushes that clean and coat it with layers of different chemicals.

First a stream of cleansing solution is sprayed on the glass, and vibrating brushes give it a good scrubbing. The glass then moves under water sprays that rinse it.

It passes under another set of nozzles that spray tin chloride on it.

Tin chloride is an undercoating that makes the next layer of chemicals stick to the glass.

The application of silver nitrate (silver dissolved in nitric acid) is the most important step in mirror making. Silver nitrate actually makes the mirror reflect. Nozzles travel back and forth above the glass, evenly spraying it. Because silver tarnishes, copper is then sprayed on top of the silver to protect it.

A coat of paint is put on to protect all the layers, and the mirror goes through a long tunnel oven that bakes on the paint. The face of the mirror is cleaned in an acid bath, and the mirror is complete.

BOTTLES

Small glass bottles have been found in Egyptian tombs dating back over three thousand years. It is believed that these bottles were made by putting glass on a bottle-shaped core made of clay or limestone and then chipping away the core after the glass had hardened. Glass making became an art about two thousand years ago when the blowpipe was developed.

The art of glassblowing has changed very little over the last two thousand years. A long iron blowpipe with a pear-shaped end is dipped into melted glass. A gob of it sticks to the end, and the craftsman blows into the pipe, causing the glass to form a bubble. This glass bubble is very flexible —it can be twirled, stretched, or squeezed into many shapes. When the glassmaker has created the shape he wants, the warm glass is cut off at the end of the pipe. The glassblower can also blow the glass into molds which shape it.

Glassblowing by hand is now used mostly in the making of decorative objects. Today, most glass bottles and other containers are made by machine.

In making glass bottles, the melted glass flows from the furnace into a dispensing machine. As the molten glass comes out of the dispenser,

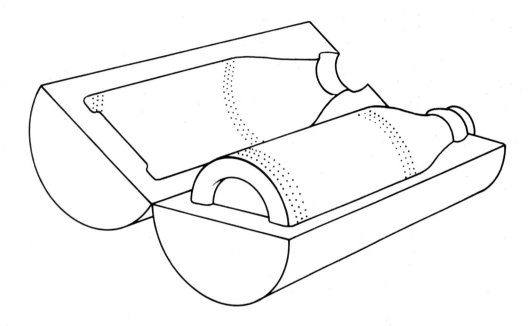

automated steel shears cut it into what are called gobs. These gobs fall, one after another, into cast-iron or steel molds, each of which is shaped like the bottle.

An automatic plunger and/or a jet of air presses the gobs down and forces them against the sides of the molds. The molds open and out come the bottles. From gob to bottle, the process takes thirteen seconds.

The bottles go on a conveyor and into a special oven called a lehr, which reheats the glass and then cools it slowly to keep it from cracking or breaking. The bottles go in the hot end of the lehr and, still on the conveyor, come out the cool end.

One glass company estimates that every person in this country uses almost four hundred bottles and jars a year.

MARBLES

People have been playing with marbles for thousands of years. Small balls of clay and stone have been found in Egyptian tombs. We usually associate the game with children, but George Washington, Abraham Lincoln, and Thomas Jefferson enjoyed playing marbles. Andrew Johnson, President Lincoln's vice president, was playing marbles when he received the news that Lincoln had been assassinated.

Today the game of marbles is universal. In some countries it is played with balls of wood, in others with balls of clay. Aggies, the most valuable marbles, are made only in Europe from agate, a semi-precious stone, although glass marbles resembling agate are made in the United States and are also called aggies.

Glass marbles are most common in the United States. To make them, a mixture of sand, soda lime, and cullet—that is, scrap glass—is heated and melted in a large tank. When molten, it is transferred to a flow tank. This glass will form the body of the marble. A

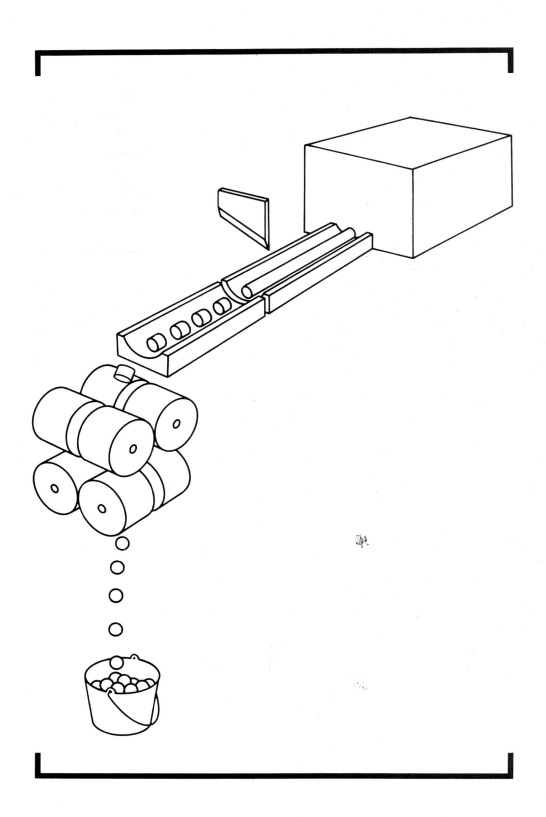

colored stream of molten glass is shot through an opening in the tank to produce striped, rainbow, or cat's-eye marbles. This striped, molten glass is then automatically dispensed from the tank as individual gobs of glass. The gobs slide down metal ramps between grooved rollers that shape them into marbles. The hot marbles roll along another metal slide, where they are sorted. The imperfect marbles are collected and returned to the heating tank to be remelted.

The remaining marbles cool in five-gallon pails that hold about five thousand marbles. From the making of the glass to the cooling of the finished marbles takes almost a day.

About 500 million marbles are made in this country each year, and additional marbles are imported. Although we usually think of them as toys, they have many other uses. They are used as "mixers" in aerosol cans. Some road signs are made of marbles imbedded in sheet metal to form numbers, words, and directional arrows. Marbles are even used in mausoleums to roll caskets into place.

METAL

Although steel was not widely produced until the eighteenth century, there is evidence that it was produced in small quantities in ancient times. Steel pegs have been found holding the stone blocks of the Parthenon in place. In the Middle Ages, the Crusaders admired the steel swords made by the Syrians of Damascus.

Steel is a combination of iron with a small amount of carbon, usually less than 1 percent. The iron used for steel is mined from the earth. Because it is not usable in this state, it is treated, processed, and then combined with other materials in a blast furnace. This iron, known as pig iron, contains 4 to 5 percent carbon. The making of steel, then, is really the removing of the extra carbon from the iron.

There are several different ways steel is made, but one of the major methods is in a basic oxygen furnace. The furnace is shaped like a pear, with a round bottom and a narrow top, and built so that it can tip, tilt, and turn. It is made of steel on the outside, but it is lined with materials that will resist heat. It can make batches of steel weighing 60,000 pounds in forty-five minutes.

To make steel, the furnace is tilted, and a load of scrap steel is dumped in by a charger, which is a cart that looks like a railroad car and works like a dump truck. Then hot, molten iron is poured in by a giant ladle held by an overhead crane. This iron melts the scrap metal.

The furnace is turned up straight, and a tube called an oxygen lance is lowered into place above the metal. Pure oxygen is blasted at high speed through the tube. It blows directly on the molten metal and burns away some of the carbon and impurities in the metal, converting it into steel.

Two minerals, lime and fluorite, are added. These are lighter than the molten steel, and they float on top of it collecting the unwanted materials and carrying them off as slag. This slag will later be used to make cement, for road fills, and in other areas of construction.

Then the oxygen is shut off, and the tube is pulled back. The steel is tested for temperature and for the correct amount of carbon. Then the furnace is tilted, and the molten steel is poured out into a ladle, leaving behind the slag. At this point other metals such as nickel or chromium may be added to create different kinds of steel, such as stainless steel, which is tougher than regular steel and resists corrosion better.

The steel can now be poured into molds called ingots. When the ingots are solid, they can be rolled into various shapes or the molten steel can be formed into solid steel in a strand casting machine. To do this, a railcar carries the ladle to an overhead crane that picks it up and carries it to the strand casting machine. The steel hardens as it passes through a water-cooled vertical mold. The steel is drawn through rolls at the bottom and then is cut by torches.

The steel is now ready to be used to make metal strips, bars, rods, wires, and other products.

LICENSE PLATES

In some states, license plates are made by the people in state penitentiaries, in others by private manufacturers.

In New York State, the main materials used in license plates are steel and reflective tape. Steel coils, weighing about a ton, are straightened and cleaned. The tape is applied to the steel. The taped steel is then cut into rectangles. These plate blanks are stored two days to allow the tape to set, so that it stays bonded to the steel.

Now the plate blanks are ready to get their numbers. Plates are made two at a time—one for the front and one for the back of the car. A pair of blanks is fed into a huge machine that is really a press. Then the dies, or stamps, are placed in the press in the proper order. These dies are in the shapes of numbers and letters and are put into the machine backward. The machine presses them into the *back* of the plate. When the plate is turned over, the letters and numbers are in the proper order and stand out from the rest of the plate. After the operator makes one pair of plates he must then change the dies for the next. An inspector checks the plates to make sure that no pair has been duplicated. One machine can make one thousand pairs per hour.

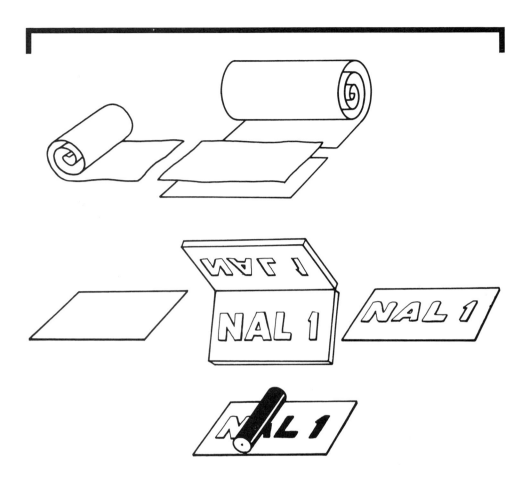

The plates are then hung on a conveyor system, which carries them to an area where they are varnished. This protects them from weather and from scratches. Still on the conveyor, they pass through a long, tunnel-shaped oven that bakes on the coating.

The plates are then fed into painting machines that apply paint to the numbers and letters by rollers.

On another conveyor they go through the oven again and the paint is baked on. When the plates come out, they are inspected. They are then packed in boxes and shipped to the Department of Motor Vehicles where they are issued.

In most states, instead of being given new plates each year, drivers are given stickers to put on their old plates, thus saving the expense of making new plates every year.

SAFETY PINS

The safety pin, that very ordinary item we depend on when we suddenly rip a seam or pop a button, is now being made of expensive materials and used as jewelry. Safety pins as jewelry may sound like the latest fad, but highly decorated pins with coil springs have been found in Egyptian and Roman tombs dating back three thousand years. These pins were mainly ornamental; for thousands of years people relied on straight pins, first made of fishbones and then of various metals, to hold things together. In 1849, Walter Hunt of New York patented a design for the safety pin as we know it today.

Safety pins are made very quickly by machines. Steel wire, loaded on spools, is cut into specific lengths, depending on the size of the safety pin desired. The wire pieces are straightened, and conveyors carry them past a grinding wheel, which makes and sharpens the point. Another wheel polishes the pieces.

Then they are fed into the assembly machine, which makes the coil in the middle of the pin and bends the unsharpened end so that it will hold the cap.

The caps are also made in the machine. They are punched out

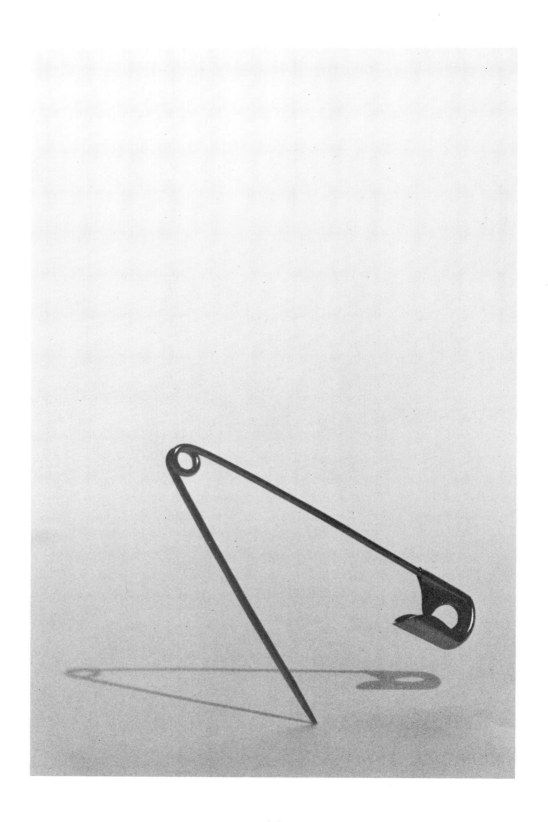

96.

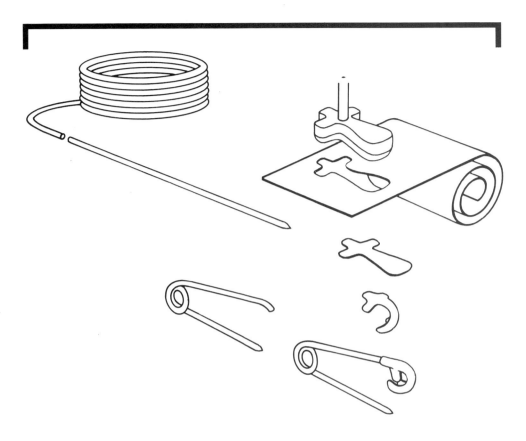

of strips of metal and shaped. The cap is then pressed on and is crimped around the bent metal.

At this time the pins are open. Some companies prefer to buy them open and ready to use, but the machine is designed to close them if necessary.

The pins are then plated electrically in a nickel solution. They are placed in a wooden barrel which is placed in a tank containing the nickel. The tank is then charged with electricity, which electroplates the pins. A gear at the bottom of the tank causes the barrel to shake and rotate so that all the pins are completely plated.

The barrel of pins is taken out of the nickel bath and put into a vat of water to clean the pins, and from there it goes into a chemical bath, which polishes the pins.

A safety-pin machine can make a pin in a minute, but it takes several hours to nickel-plate, polish, and rinse them. Over 100 million safety pins are made a year; they seldom have a chance to wear out—they are usually lost or misplaced.

SPOONS

Most of the knives, forks, and spoons we eat with every day are made of stainless steel. Stainless steel has been used in the making of flatware—knives, forks, and spoons—for only about forty years, and it has only been in the last twenty years that ways have been discovered to make patterns on the stainless steel and to make it in different designs.

Stainless steel is popular because it does not tarnish or scratch easily. It is a combination of steel with two other metals, chromium and nickel. Steel is scratch-resistant, chromium keeps the metal "stainless," and nickel gives it its sheen and brightness.

Stainless steel comes from the mills in sheets or coils. To make a spoon, the stainless is fed into a machine which cuts a spade-shaped rough outline or blank of the spoon. This blank is somewhat smaller than the finished spoon. The bowl and handle parts are then mechanically rolled out to the desired thickness and proper length.

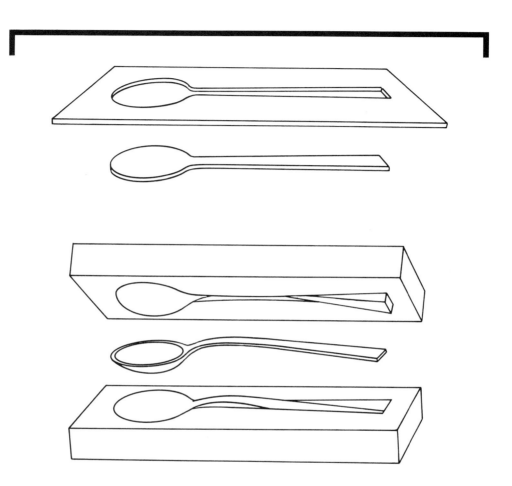

This rolled flat blank goes to another machine, which cuts out the spoon form and takes off the extra metal from the outside edges.

The blank is heated to make it soft and placed on a steel die or mold. This mold is the exact shape of the back of the spoon. Another steel die, which shapes the front of the spoon, is mounted in a drop hammer that weighs about 1,000 pounds. With tremendous force, the top die drops twenty inches onto the spoon blank, hollowing out the bowl and curving the handle.

The design is put on the spoon by dies which have the design embossed in reverse on them. Because the design stands out, when the top die slams down on the spoon, a perfect pattern results.

Other machines are used to trim, wash, and polish the spoon. After a final polishing by hand to give it a bright shine, the spoon is inspected. It is then ready to be packed, bought, and used.

WHISTLES

There are many kinds of whistles for many different uses. There are toy whistles, dog and bird whistles, scout whistles, and police whistles. The best, such as those that policemen use, are made of solid brass.

Large coils or spools of brass are unrolled and go directly into a die press that operates like a big cookie cutter. It cuts the brass into three different flat shapes or blanks: the outer body, which is shaped like a butterfly and will form the sides and the mouthpiece; the outer coil, which will form the back and the part above the air hole; and the tongue, which is inside and directs the air in, around, and out the open hole at the top of the whistle.

Another press bends the blanks into the right shapes. These formed pieces are fitted together by hand. One man can put together about a thousand whistles a day.

The parts are then soldered or fused together, cleaned, and pol-

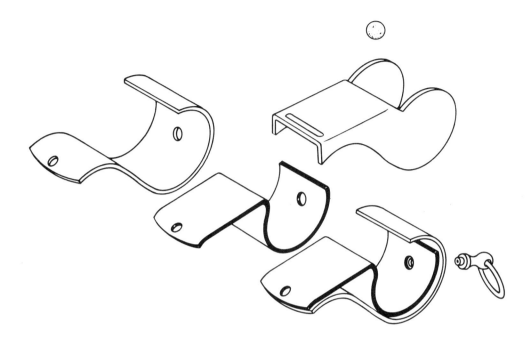

ished. This can be done either by hand or by machine. The whistles are put on racks and dipped into metal baths of liquid nickel, silver, or gold. The solution is charged with electricity, causing the liquid metal to plate the whistles, and the whistles are then rinsed in water.

Next the ball is inserted in the whistle, giving it its trill. The ball is made of cork and is put into the whistle by hand. It is squeezed and flattened until it is pushed through the hole. Once inside, the ball bounces back to its original shape.

Plastic lip guards are often put on police whistles so the lips won't stick to the cold metal in winter.

Finally the whistles are tested for sound. There is actually a whistle tester who listens to the sound of each whistle. He puts the whistle in front of an air compressor which sends a blast of air through the mouthpiece. The whistle tester listens and approves or rejects a whistle by its sound.

PENNIES

When the English colonists came to America, they used their own currency. The first mint in America was begun in 1652 in Boston, and this mint made shillings, sixpences, and three-penny pieces.

In 1785 the United States government approved a resolution to base its monetary system on the dollar and to adopt the decimal system. The first authorized coin was a copper cent. On it was the Latin word *fugio*, meaning "I fly," which appeared above a picture of the sun. Also on it were the words "Mind your business." George Washington refused to let his portrait be used on coins; he didn't think it was a democratic thing to do.

The correct name for a penny is a cent. A cent is one-hundredth of a dollar; the word "penny" comes from the name of an English coin which is considerably larger than our cent.

All coins in this country are made by the Bureau of the Mint in three places—San Francisco, Denver, and Philadelphia. More pennies are made than any other coin, about 9 billion a year.

Pennies are 95 percent copper and 5 percent zinc. The metals are heated and melted together and cooled in a mold. The molded metal passes

through many sets of rollers to make it thin, blanks shaped like pennies are stamped out of the metal, and the design is pressed into them.

To start at the beginning, the metal is placed in a container and weighed. A crane picks up the box and carries it to an electric melting furnace that holds 15,000 pounds of metal.

The melted metal is poured into giant molds, forming ingots which look like huge candy bars. Each ingot is 6 inches thick and about 18 feet long, and weighs about 7,000 pounds.

After cooling, an ingot goes by conveyor to a circular saw, which cuts it into two slabs of equal weight. Each slab is reheated until soft. When it becomes red-hot it is squeezed through several rollers until it becomes a strip only half an inch thick. Two sprays of water then cool it.

A machine shaves the top and bottom of the strip to make it shiny and smooth. The shavings go back to the furnace so no metal is wasted.

Even though the strip is so thin it can be rolled, it must be made thinner. It goes through two more rolling mills until it is only one-twentieth of an inch thick.

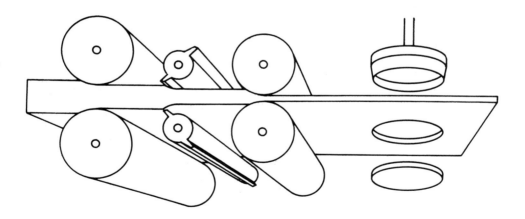

This thin strip goes to the blanking press, which is like a cookie cutter. It punches out round pieces that are slightly larger than a penny.

These blanks are put in another furnace, where they are softened. They are then cooled in a water bath. This heating and cooling tempers, or strengthens, the metal. These cold blanks are then washed in a big washing machine filled with hot detergent. The detergent polishes and cleans them. The blanks are loaded into a dryer and tumbled until dry.

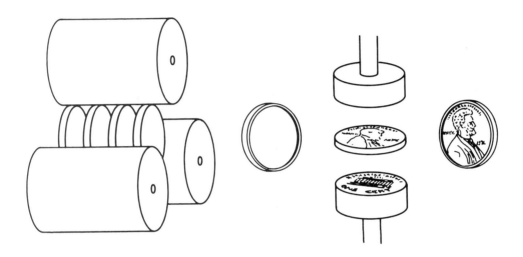

The pennies are rolled on their edges through a machine that applies enough pressure to raise a rim around the blanks.

At last the blanks are ready for their designs—President Lincoln's portrait on one side and the Lincoln Memorial on the other. The designs, front and back, are engraved on steel stamping devices in a press that works like a waffle iron. The blanks are put between the dies and "fingers" on the press firmly hold each blank. The top die comes down and with one heavy blow the design is stamped on each side.

After inspection, defective pennies will be melted again, and the good ones are counted in a machine that then drops them into a large bag. After exactly 5,000 pennies have fallen into the bag, it is automatically sewn closed.

The bags of pennies go to the twelve Federal Reserve Banks and their branches, which distribute them to banks throughout the country.

FOODS

COCOA

ocoa is made from cacao beans. The name of the genus to which the cacao tree belongs is *Theobroma*, "food of the gods." The Mexican Aztecs believed that the cacao seed was a gift from God, brought to earth from Paradise by Quetzalcoatl, the god of air.

The Aztecs used the cacao beans to make a drink called *chocolatl* (the word meant "warm beverage"). The cacao beans were dried in the sun, roasted in earthen pots, and then ground between stones. The result was a thick paste that the Aztecs mixed with water, then molded into cakes. When they wanted *chocolatl*, they pounded the cakes into powder and mixed it with water. Their drink was very rich, hard to digest, and thick as honey. It had to be dissolved in the mouth before it could be swallowed.

Today, the making of cocoa is more complicated, involving cleaning, hulling, and blending the beans, but the beans are still dried, roasted, made into cakes, and ground.

Cacao beans grow in pods, 6 to 10 inches long, on the trunk and branches of the tree. The pods resemble gourds. The trees grow in warm, moist climates. Africa produces 75 percent of the world supply of cacao. Experienced workers use long-handled knives to cut off the pods,

being careful not to injure the tree. Then, with machetes, they cut open the pods. The beans, which are covered with pulp, are scooped out by hand. The pulp-covered beans are put on the ground and covered with plaintain or palm leaves. This process browns the beans and hardens the outside of them into a shell that will be taken off later in the factory.

Then the beans are dried. They can be spread out to dry in the sun on palm leaves, which can be rolled up in case of rain and carried indoors. Or they can be placed on wooden trays that have removable roofs. Either method takes a week or two. They can also be dried more quickly in mechanical dryers.

The dried beans are transported to a factory where they are cleaned, roasted, hulled, blended, and ground.

First the beans are cleaned by moving over a number of vibrating sieves. Strong air jets blow away lightweight material.

The beans ride conveyors to the roaster, a revolving drum in which streams of heated air cook the beans.

Still in their shells, the beans go through rollers that crack them into fragments called nibs. This is the "meat" of the bean. The nibs are separated from the shells by going through more sieves. Air blows shell fragments away from the heavier nibs.

The nibs are then ground in milling machines, which consist of several sets of grinding plates that revolve rapidly. This friction causes heat, which breaks down the nib cells, forming a free-flowing mass, chocolate liquor. This is the basic ingredient of all chocolate, including chocolate bars. Some of this liquor is cooled, hardened, and sold as baking chocolate.

Chocolate liquor can also be separated into cocoa butter and cocoa powder in a press. The press squeezes out the butter, which will be used in chocolate candy bars and suntan lotions. The cocoa powder leaves the press as a solid cake of cocoa. The cakes are crushed between rollers until they are a soft brown powder. Some cocoa manufacturers add sugar to the cocoa before packaging it, so that all you have to do is add milk or water to make a cup of cocoa. Others let you add both the sugar and liquid at home.

CHOCOLATE BARS

Chocolate is one of the most popular foods. We eat it in cakes, cookies, and ice cream, use chocolate syrup in milk, and eat it as candy in the shape of bars, kisses, and drops.

Step by step, this is how chocolate bars are made. Sugar and milk are combined in large kettles. Most of the water from the milk is boiled off, leaving a taffylike substance. This is placed in a big mixer. Chocolate liquor is added, as well as some cocoa butter (discussed in the previous chapter). It becomes pasty, and is ground between steel rollers to make it smoother.

It then flows into a conch machine, so named because it is shaped like a conch shell. For several days the chocolate is kneaded, pushed, and turned under granite rollers. Finally it is velvety smooth.

After conching, the melted chocolate is put into a "tempering kettle," where it is reheated and then cooled. Tempering helps keep chocolate fresh after it is packaged.

At last the chocolate is ready to be made into bars. It is poured automatically into rows of metal molds, which move on a conveyor through a cooling tunnel.

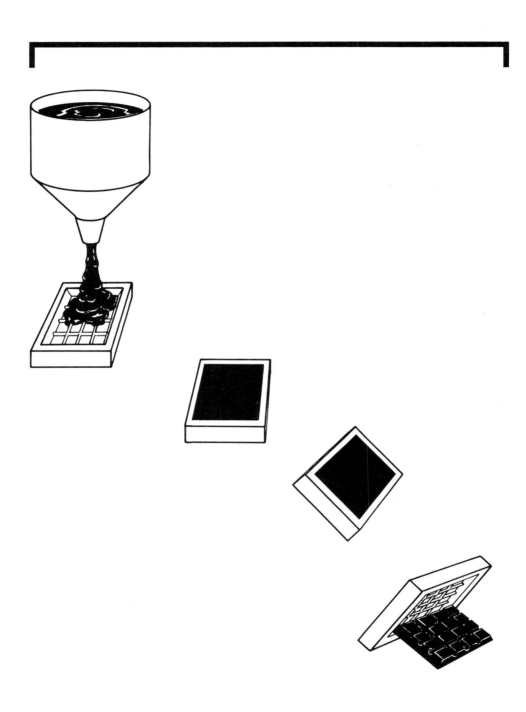

After they are cool, the molds are turned upside down and the bars fall. They are then wrapped with a protective waxy paper, and a printed paper which identifies the chocolate bar and makes it look attractive.

HOT DOGS

There is a story that the hot dog got its name in 1900 when a man named Harry Stevens, owner of the refreshment stands at the New York Polo Grounds, was looking for a hot food to sell to football fans. He decided to sell sausages. The vendors went around shouting, "Get your red-hot dachshund sausages!" That gave sports cartoonist Tad Dorgan the idea of creating a talking sausage as a cartoon character. It has been said he didn't know how to spell "dachshund," so he called it "hot dog."

It may not be a true story, but it is certain that the hot dog is one of the most popular American foods. About 14 billion are eaten every year.

Hot dogs are made from hog and/or beef trimmings, although some are made from chicken. The manufacture of hot dogs is totally automated. Chunks of meat are put in a grinding machine. The ground meat goes into a huge mixer which can hold 9,000

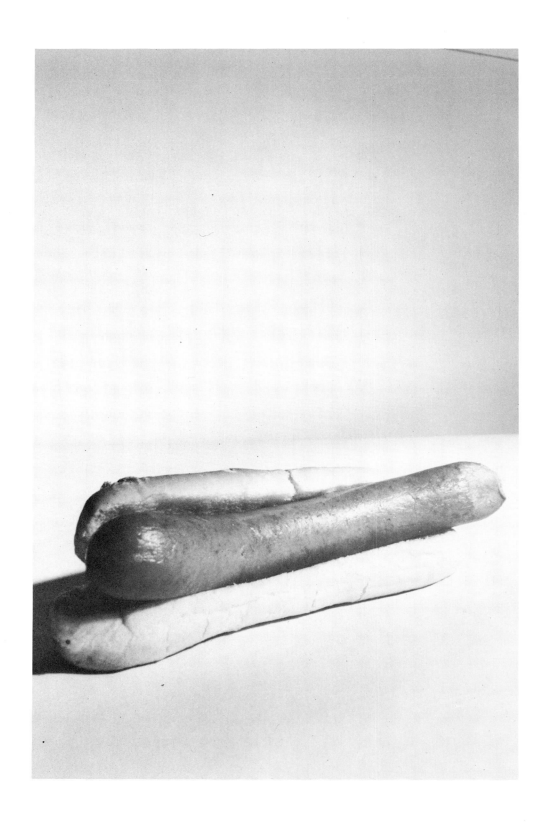

pounds of meat. The meat is mixed with salt, sugar, spices such as ground mustard and nutmeg, water, and a preservative, sodium nitrate. It now looks like cake batter. This batter is pumped into a chopper that grinds it up even more.

Another pump stuffs the batter into 110-foot-long cellulose casings. This casing acts like a cake pan, keeping the batter in the right shape until it is cooked. These long tubes go on a conveyor through a linking machine that pinches them into regular hot-dog-length sections.

They continue on their conveyor trip through the "smoking" room, where, in a few minutes, enough smoke seeps into the hot dogs to flavor them. Then they go through a series of heated cabinets that completely cook them. Still traveling along, they pass through a chilling tunnel and right into a machine that slits the cellulose and removes it from the cooked hot dogs.

The hot dogs continue on to a packing device, which automatically picks up the hot dogs and packages them.

The entire process, from raw meat to cooked hot dogs, takes one hour. In that one hour, 36,000 hot dogs can be made without human contact.

ICE CREAM

About a billion gallons of ice cream are eaten every year. Vanilla, chocolate, and strawberry, in that order, are the most popular flavors.

Although about 47 million Americans eat ice cream every day, few of them know that ice cream has had a colorful history.

The Roman Emperor Nero sent his slaves to the mountains to bring back snow and ice, which were then flavored with honey and fruit pulps and served to him. This was the ancestor of ice cream.

Marco Polo brought a recipe for sherbet to Europe from the Far East.

In the seventeenth century, a French chef served Charles I of England a "frozen milk" dessert at a royal banquet. The king was so pleased that he paid the chef a large sum of money every year to keep the dessert for the royal table alone. Fortunately for ice-cream lovers, the chef did not keep the secret to himself.

The colonists brought ice cream to the United States, and by the early nineteenth century, ice cream was being served in coffee houses and ice-cream parlors.

121.

Ice cream is made by combining milk, cream, sugar, and flavorings. The mixture is heated and then whipped while it is freezing. This simple process can be done at home if you have the proper equipment.

In the ice-cream factory, however, the same process is directed by complex technology from a special control room. All the ingredients to make ice cream are stored in huge tanks. These tanks are connected to the control room, where a few people can make enormous quantities of ice cream just by pushing buttons and adjusting dials.

The cream and milk solids and sugar flow into a mixing tank, where they are carefully blended. The mixture then flows through tubes and is pasteurized—heated at a high enough temperature to kill harmful bacteria.

Then the mix is homogenized by being forced through a small valve under high pressure. Homogenization means that the milk fat is broken up into tiny pieces and spread evenly through the mix, thus making the ice cream smooth.

The mix flows into a cooler and from there into a freezer. Fruits, nuts, and flavorings are added by an automatic "flavor feeder." While the ice cream is being frozen, blades in the freezer whip the mix. Without the whipping, the product would be just a hard mass of cream, milk, sugar, and flavoring. This whipping, or "overrun" as it is called, adds air to the mix, making it light and smooth. To ice-cream experts there is a fine balance between too much air, which makes light, weak-tasting ice cream, and too little air, yielding dense, hard ice cream. So it's not just the ingredients but the overrun as well that makes one brand taste better than another.

The ice cream is at a partial or soft-frozen stage when it is piped from the freezers to machines that package it into containers, paper cups, or molds. The packaged ice cream moves on conveyors to the hardening room, which is a huge refrigerator with a temperature below zero. It is kept there until it is ready to be loaded onto refrigerated trucks and shipped to the stores.

POTATO CHIPS

Potato chips are a very popular snack —the average person eats more than 4 pounds each year.

Potato chips can be made at home as well as in a factory. The only difference is the amounts made and who does the work. In either case the potatoes are sliced very thin and deep-fried in cooking oil, then salted.

Potato-chip companies "age" their potatoes in special storage rooms before they use them. That's because potatoes have sugar in them. As they get older, the sugar turns to starch, which gives potato chips their special color and flavor. Those dark-brown potato chips that we sometimes find among the others are not really burned. They are chips from a potato that had too much sugar in it. The more sugar, the darker the potato chip.

The aged potatoes go into a hopper, which is really a giant funnel. It feeds the potatoes into a brush washer, where the dirt and sand are washed and cleaned off the potatoes.

The potatoes then travel through a heavy stream of water. They bob along through the water, and any stones that might have been picked up in the potato fields drop down to the bottom.

Next the potatoes go into a continuous automatic peeling machine

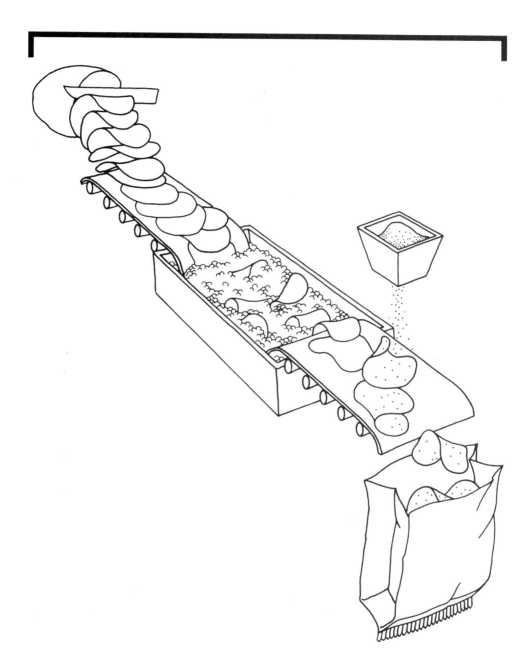

that rubs off the peel. Then they ride on a moving belt past inspectors, who trim out spots, clean out the deep eyes, and get rid of any bad potatoes. This is the only time the potatoes are touched by human hands. Everything else is automatic.

They travel on to the potato slicer. This is a large whirling tub with eight razor-sharp slicing blades. The potatoes are thrown against blades that cut off eight slices at a time. Twenty slices are cut from an inch of potato.

Because the slicing releases starch and juices, the slices are washed and rinsed. They are then spread in a thin layer on a wide conveyor, and they go into the potato-chip fryer. This is a long kettle with 6 inches of cooking oil in it. The potato slices enter at one end of the kettle, spend three or four minutes moving through the hot cooking oil, and come out the other side as potato chips. They pass under an automatic salter which sprinkles salt on them.

Continuing on their conveyor journey, they are inspected. Burned or broken chips are picked out. Now the chips go on an overhead conveyor, where they are fed into package-filling machines.

These machines do everything. They form the bags, weigh the chips, seal the bags, and send them off on another conveyor, which takes them to the area where they are packed into cases for shipping. It takes less than thirty minutes to turn a batch of potatoes into bags of potato chips.

CORN FLAKES

The Kellogg Company in Battle Creek, Michigan, introduced its first product, corn flakes, to the American public in 1906. And it has been one of the most popular cereals ever since.

The discovery of the corn flake was actually an accident. Will Keith Kellogg was an employee of the Battle Creek Sanitarium, where his brother, Dr. John Harvey Kellogg, was the chief surgeon and superintendent. They were both doing experiments with food to find tasty ways to serve their patients vegetarian meals, which would be easy to digest. While experimenting with a new kind of bread, they placed some boiled wheat berries on a baking pan in the oven. They were both called away from the laboratory, and when they returned the baked wheat berries had dried. When they rolled the berries, each one flattened into a large thick flake. They had stumbled onto a method for making wheat flakes. By the same method they were able to make corn flakes, rice flakes, and other similar breakfast foods. When their patients left for home, they placed orders for these breakfast foods with the sanitarium. Will Kellogg, a sixth-grade drop-out, went on to form the company that makes cereals.

Corn flakes are made today very much as they were then. Corn

grows from a kernel to maturity in four months. The corn is harvested by a machine called a picker-sheller, which does what the name implies; it picks, shells, and slices the corn kernels off the cob. They are brought to a corn mill, where they're steamed. This process loosens the hull from the endosperm, or meat, of the corn. The endosperms, also called grits, are dried after being separated from the hulls.

The grits are then cooked with malt, sugar, and salt in giant stainless-steel rotating cookers under steam pressure. This works in the same way as a pressure cooker. The cooked grits are collected in holding tanks, where they are dried and tempered or seasoned.

The grits are flaked in a flaking mill, made up of a pair of large, heavy steel cylinders. The smooth rollers rotate in opposite directions. Since the back roller is hotter than the front roller, the flakes stick to the back roller. This process stretches as well as flattens the flake.

The flakes are tumble-toasted in giant rotary ovens at high temperatures. After the cereal is toasted, it travels along a conveyor belt, where it is sprayed with a vitamin solution and inspected.

After inspection, the corn flakes move to the packaging area and are weighed and transferred into cartons lined with wax paper or aluminum foil. Packed in cases, the corn flakes are shipped by road and rail and spoon to their destination.

OTHER PRODUCTS

BASEBALLS

Baseball has been called the "national pastime" of the United States, for it is played by everyone from third-graders to major-league players.

For any baseball game, the type of ball used is very important. From the inside out, this is what the best balls are made of.

The center of a professional ball is a small, perfectly round ball of cork and rubber and is called the "pill." If only rubber were used, the ball would have too much bounce. The pill is surrounded by two rubber half-shells with cushioned washers between the two shells. This is all covered by soft rubber, which in turn is covered by various yarns. Finally, the yarns are covered by white leather.

The entire center—pill, rubber halves, rubber covering—is put together by machine and shipped to the factory which actually makes the balls.

At the baseball factory, these centers are covered by five different windings of wool and cotton yarns. The winding is done by specially designed machines. For each of the first two windings, a four-strand gray wool yarn is used. A third machine winds on three-strand white virgin wool yarn. The fourth winding is three-strand gray yarn, and the fifth is a fine two-

strand white cotton yarn. This last yarn gives a firm, smooth surface so that the cover can be applied.

Then the ball is dipped into rubber cement to seal and bind the yarn.

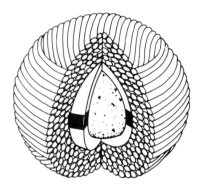

After each step, the ball is inspected. After the dipping, the ball is tested for size and weight. It must weigh between 5 and 5¼ ounces and measure 9 to 9¼ inches around, when it is finished.

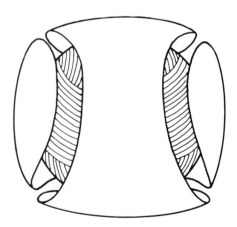

The ball is now ready for its cover. The covers are cut by machine from white leather. The covers are then split on a splitting machine to the correct thickness and weight. The cover pieces are matched in pairs.

Up to this point, everything has been done by machine. The covers, though, are sewn on by hand. No one has been able yet to invent a machine that can make the right stitches. It is necessary to make sure the first and last stitches are tucked under in such a way that they are held securely and don't make a bumpy seam. Experienced sewers put the two pieces of leather around the ball, put the ball in a vise, and, using thick twine and two needles, sew the cover on the ball.

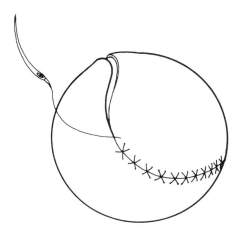

After the balls are stitched, they are put between two hard maple circular rollers which press and flatten the seams. At the same time, a fine powder is put on the balls to polish the covers.

Every ball is weighed, measured, and carefully inspected. Then the balls are stamped with the proper identifying information, so that the consumer will know what type and quality of ball he is buying. Machines wrap and package the balls, and they are ready to be used.

CRAYONS

We think of crayons as a drawing and coloring "tool" for children. Yet children have been using them for only about seventy years. Before that, crayons were very expensive and were owned and used by artists alone.

The first crayons, which were used in the fifteenth century, were a mixture of charcoal and oil. Later, colored powders called pigments were added, and it was discovered that if wax was added, the crayons were sturdier and easier to handle.

Crayons were first made in this country in 1903. They were made much the same way then as they are today; heated paraffin wax was mixed with colored pigments, mixed, "cooked," and molded into sixteen different-colored crayons.

Today, crayons come in as many as sixty-four different colors. Crayon makers follow an exact "recipe" for making them so that the colors will always be the same from batch to batch.

At the crayon factory, the paraffin wax, in liquid form, is stored in huge outdoor heated tanks. Each tank is 26 feet high and can hold 17,000 gallons of wax. Attached to these tanks are a number of pipelines, which are connected directly to a giant heated mixing vat.

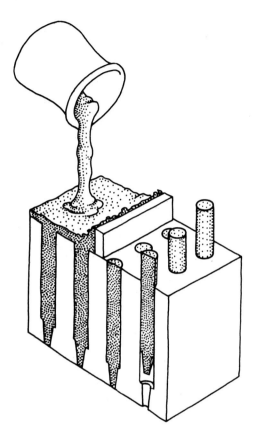

When ready to make crayons, the "mix-man" turns on a faucet in the vat and lets the liquid wax run into the mixing kettle. The pigment, any one of the sixty-four colors, is added. It must be carefully measured to get the exact color desired.

Stearic acid, an animal fat, is added to make the wax strong enough so that the crayons won't soften when they are held and to make them smooth and easy to draw with.

The mix-man lowers large mixer blades into the kettle and turns them on. After the wax and pigment are thoroughly mixed, the "molder" draws the colored wax into a bucket and carries it to a nearby molding machine containing over two thousand molds. This is really a big metal table full of holes. He pours the hot wax over the table, and it flows down into the holes, each of which is a mold the same shape and size as a finished crayon. It is then allowed to cool and harden. The extra wax on top of the table is scraped off and returned to the mixing kettle. Now the round bottom ends of the crayons can be seen. A special device automatically pushes up all the crayons out of their molds at once and right into trays called "holding racks."

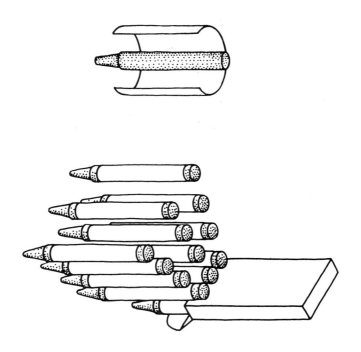

The racks are taken to an inspection table and carefully emptied, making a neatly stacked pile. An inspector checks the color and shape of the crayons.

After they are approved, they are fed into a labeling machine, which puts wrappers around each of them. Then the crayons go into packaging machines. Since boxes can hold from eight to sixty-four crayons, there are different packaging machines for the differing sizes. In the machines, each color goes into a separate compartment and is fed out into a main compartment. In that way, each box gets one of each color.

CHEWING GUM

People have been chewing gum for over a thousand years. The Mayans, a tribe of South American Indians, chewed chicle, a sticky substance which comes from the sapodilla tree. The American Indians of New England chewed a gum made from the resin of spruce trees. But it was not until the late nineteenth century that chicle was brought to this country. The Mexican statesman Antonio Lopez Santa Anna asked an American inventor named Thomas Adams to try to make rubber from chicle. He was not able to do it, but he discovered that the chicle was good for chewing. He made a batch of gum and talked his local druggist into selling it. That was the beginning of the many colors, flavors, and shapes of chewing gum we can buy today.

Chewing gum is made up of the base, sugar, corn syrup, and flavorings. The "base" means the latex called chicle, together with other latexes, or man-made substitutes. The ingredients are heated and mixed together, and the mixture is put through a machine that shapes the gum into large sheets. Rollers flatten the sheets, and they are dusted with sugar. The gum is allowed to "set" for a few days and it is then cut into sticks and wrapped.

The making of gum really begins with the tapping of sapodilla trees to get chicle. This is a form of latex, a milky fluid that comes from certain trees, most of which grow in tropical forests and jungles. Tires and gum have something in common—latex is used in making both of them. The latex used in chewing gum does not come from the same tree as the latex used in making rubber products. However, methods for getting this substance are almost the same. Workmen make V-shaped cuts in the bark. The chicle flows down the cuts into buckets. It is boiled to get rid of extra water, and then is shaped into blocks by hand, and the blocks are shipped to the factory.

Here the chicle is heated in large kettles. When it is hot enough, it looks like thick maple syrup. This hot "syrup" is strained through a number of fine mesh screens.

Then the base is put in a huge mixer which has slowly revolving blades. As the blades turn, corn syrup and powdered sugar are added. The corn syrup helps the sugar combine with the gum base and keeps the gum moist and pleasant to chew. Extra softeners are sometimes added, which also help keep the gum moist. Finally, flavorings are added. They may be in the form of flavoring oils, such as peppermint, or else finely ground citric or other acids from fruits for grape, orange, cherry, and other fruit-flavored gums.

When the mixing is done, the blended gum is cooled on cooling belts. Then it moves into an extruder, a machine which presses and kneads the gum to make it smoother, then pushes it out a long slit in the machine. The gum comes out in the shape of thick sheets.

The thick sheets pass into a "sheet-rolling" machine. Giant rollers flatten the gum into thinner and thinner sheets.

These sheets travel into a machine which cuts them into smaller sheets. Then the machine scores them—that is, cuts lines into them in the shape of single sticks of gum. The machine does not cut through the gum; it just makes a pattern on it.

The gum goes into another roller. When it comes out, it has also been sprinkled with powdered sugar. The roller pulls the gum onto trays. These trays of gum are put in an air-conditioned room for at least two days. Letting them "set" makes them easier to break and wrap.

When the gum has set, it is put through a breaking machine, which breaks it up into sticks and puts them into trays. These sticks are put into gum-wrapping machines, then boxed and shipped.

Ball gum and candy-coated gum are made in much the same way
—up to a point. The candy-coated gum is scored into little squares or
oblongs, then broken up by machine. The ball gum is scored into pencil
shapes and then turned into balls by machines.

The balls and candy-coated gum are allowed to set for a day or
two. Then they are put in coating pans, where they are first coated with a
liquid gum so that the sugar coating will stick. After it is covered with
pure liquid sugar, the gum is "polished" by putting it into pans where it is
whirled with beeswax.

Bubble gum is made by adding a small amount of rubber latex,
which makes the gum stronger and stretchy so that large bubbles can be blown.

SOAP

About once a year, a pioneer woman made soap. She used only two things—fat and potash. She got the fat by collecting all the drippings from the meat she cooked during the year. She got potash, which is an alkali—a kind of salt that will dissolve in water —by pouring hot water through wood ashes. In a large kettle, she mixed the potash and the fat drippings together and then boiled them outdoors over an open fire. After letting the mixture cool, she cut it into bars. The soap was rough, and it didn't smell good, but it worked.

Today, although our soap is made for us in large factories, these same two ingredients are still being used, and they are still mixed and then cooked. But soap has been improved in many ways. The fat now goes through certain processes to make it purer; deodorants, perfumes, and cleansing creams are often added; and the soap mixture is often pounded and kneaded by machines to make it lather better.

The pioneer woman's soap wasn't as good, but she made it much more simply.

Soap companies use both animal fats and coconut oil. Fats are made up of fatty acids and glycerin. Fatty acids are needed for the soap

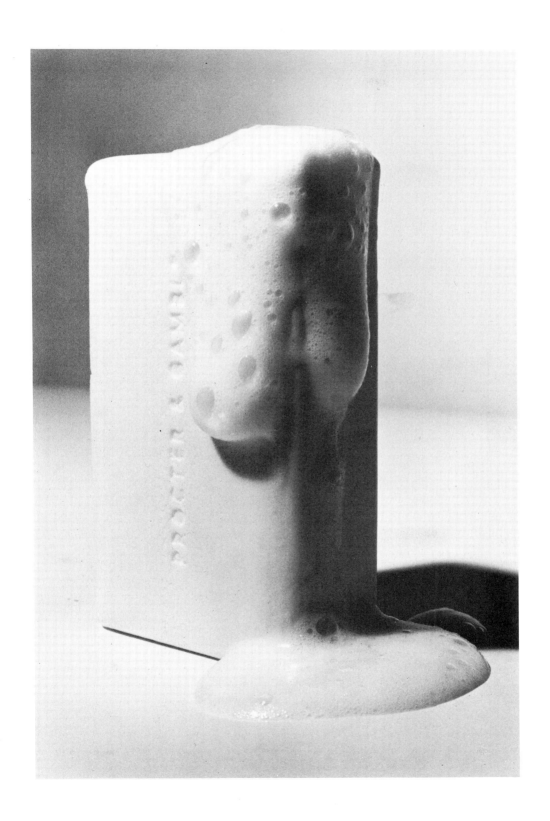

145.

and glycerin isn't, so the two must first be separated. This is done in a hydrolyzer, a stainless-steel tube as big as a barrel and 80 feet high. It breaks down the fat into fatty acids and glycerin. The fatty acids are drawn off, and they go into a "still" to make them purer.

Then the fatty acids travel into a "metering pump." Next to it is another pump containing the alkali. The right amount of each is measured out into a mixer. Once mixed, the thick liquid passes into a blender and then into another mixer where other ingredients, perfumes, and deodorants are added.

At this point, the "neat" soap, it is called, can go in two directions. If it's a floating soap, it goes into a machine called a freezer. This works like an ice-cream-making machine, whipping the mixture into a cream and cooling it at the same time. (It is the air that is whipped into it that makes it float.) When ready, it is pushed out in an endless piece of soap and then goes through a cutter which cuts it into bar sizes. The bars are stored to let them get hard, then stamped and wrapped by machines.

If the "neat" soap is to be used as a soap which will lather a lot and will work in cold water, it travels a different route. It goes from being a liquid to being soap flakes to being, at last, a bar of soap. Here's how. The neat soap goes between heavy rolls or mills which cool it and grind it up into small pieces. The pieces go through a dryer, then into another mixer, then into more mills, and finally into a machine that presses all the pieces together and then pushes out one long, continuous bar of soap.

It takes six hours to make a bar of soap.

JEANS

Jeans are considered real American clothing. Every country has its own style of work clothes, and blue denim was once unique to the United States. Though jeans were designed for their durability, their popularity has spread all over the world and they are worn in the most fashionable places. Once they were forbidden in elementary and high schools; today, they are almost a uniform in some parts of the country.

The word "jean" comes from the name of a coarse fabric that was made in Genoa. "Denim" comes from a fabric made in Nîmes, a town in southern France, and called *serge de Nîmes*.

Standard blue jeans are made from a stock design that hasn't changed over the years. Designers have added a variety of styles to the basic work jean, with flared pant legs, wide belt loops, patch pockets, and all kinds of trims. However, all jeans use the same basic raw materials—fabrics, thread, buttons, snaps, and zippers.

From the design, a pattern is made. They are made from heavy, cardboard-like material, and every size of a particular style jean will have its own set of pattern pieces. There are about ten different pieces for each pair of jeans.

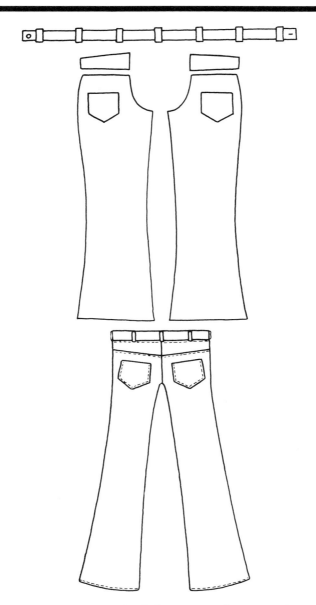

The fabrics are unrolled flat onto a large cutting table. The patterns are arranged on the fabric, and high-speed cutting machines cut out many layers of fabric at one time.

After the fabric is cut, the small pieces are gathered together in bundles, ready to be assembled.

Most of the work in making jeans is sewing the pieces of fabric together. Each step is done by a different sewing machine operated by a

149.

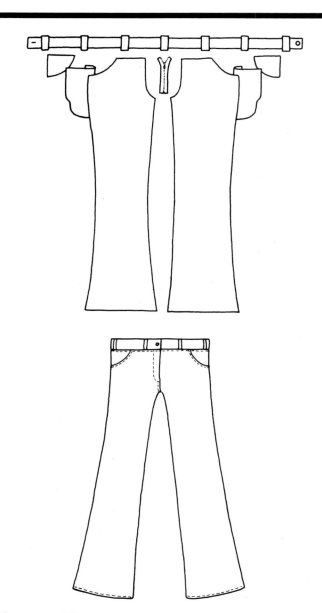

highly trained person. The sewing operation is really a series of single jobs such as putting the pockets together, making belt loops, and attaching the pockets. Still others attach the belt loops, sew the pant legs, attach the buttons, sew the hem, add the zipper, and finally add the label.

The assembled jeans are then pressed in big steam presses and tags are attached to identify the size, and they're ready to go to stores all over the world.

Danelle McCafferty is the author of *Celebration: The Wild Flower, Write-Your-Own-Ceremony, Picnic Reception Wedding Book*. She has done public relations writing, copywriting, and ghost writing and has conducted in-depth interviews for a study on fatherless families.